Collected Works of A.M. Turing

MORPHOGENESIS

Collected Works of A.M. Turing

Mechanical Intelligence
Edited by D.C. INCE

Pure Mathematics
Edited by J.L. BRITTON

Morphogenesis
Edited by P.T. SAUNDERS

Mathematical Logic
Edited by R.O. GANDY and C.E.M. YATES

NORTH-HOLLAND
AMSTERDAM · LONDON · NEW YORK · TOKYO

Collected Works of A.M. Turing

MORPHOGENESIS

Edited by

P.T. SAUNDERS
King's College, London, United Kingdom

1992
NORTH-HOLLAND
AMSTERDAM · LONDON · NEW YORK · TOKYO

ELSEVIER SCIENCE PUBLISHERS B.V.
Sara Burgerhartstraat 25
P.O. Box 211, 1000 AE Amsterdam, Netherlands

ISBN: 0 444 88486 6

Library of Congress Cataloging-in-Publication Data

Turing, Alan Mathison, 1912–1954.
 Morphogenesis / edited by P.T. Saunders.
      p.   cm. -- (Collected works of A.M. Turing)
 Includes bibliographical references and index.
 ISBN 0-444-88486-6
 1. Plant morphogenesis. 2. Plant morphogenesis--Mathematical
models. 3. Phyllotaxis. 4. Phyllotaxis--Mathematical models.
I. Saunders, P.T. (Peter Timothy), 1939–   . II. Series: Turing,
Alan Mathison, 1912–1954. Works. 1990.
QK665.T87   1992
581.4--dc20   91-34306
                                                          CIP

Acknowledgement is gratefully made to the Royal Society for permission to reprint
"The Chemical Basis of Morphogenesis in Plants", Phil. Trans. R. Soc. London B 237
(1952) 37–72.

Hockey

or

Watching the Daisies

Grow

Photostat copy of caricature by his mother of Alan at hockey; sent to Miss Danvers, matron at Hazelhurst. Date. Springterm 1923

PREFACE

It is not in dispute that A.M. Turing was one of the leading figures in twentieth-century science. The fact would have been known to the general public sooner but for the Official Secrets Act, which prevented discussion of his wartime work. At all events it is now widely known that he was, to the extent that any single person can claim to have been so, the inventor of the "computer". Indeed, with the aid of Andrew Hodges's excellent biography, *A.M. Turing: the Enigma*, even non-mathematicians like myself have some idea of how his idea of a "universal machine" arose – as a sort of byproduct of a paper answering Hilbert's *Entscheidungsproblem*. However, his work in pure mathematics and mathematical logic extended considerably further; and the work of his last years, on morphogenesis in plants, is, so one understands, also of the greatest originality and of permanent importance.

I was a friend of his and found him an extraordinarily attractive companion, and I was bitterly distressed, as all his friends were, by his tragic death – also angry at the judicial system which helped to lead to it. However, this is not the place for me to write about him personally.

I am, though, also his legal executor, and in fulfilment of my duty I have organised the present edition of his works, which is intended to include all his mature scientific writing, including a substantial quantity of unpublished material. The edition will comprise four volumes, i.e.: *Pure Mathematics*, edited by Professor J.L. Britton; *Mathematical Logic*, edited by Professor R.O. Gandy and Professor C.E.M. Yates; *Mechanical Intelligence*, edited by Professor D.C. Ince; and *Morphogenesis*, edited by Professor P.T. Saunders.

My warmest thanks are due to the editors of the volumes, to the modern archivist at King's College, Cambridge, to Dr. Arjen Sevenster and Mr. Jan Kastelein at Elsevier (North-Holland), and to Dr. Einar H. Fredriksson, who did a great deal to make this edition possible.

P.N. FURBANK

ALAN MATHISON TURING – CHRONOLOGY

1912 Born 23 June in London, son of Julius Mathison Turing of the Indian Civil Service and Ethel Sara née Stoney

1926 Enters Sherborne School

1931 Enters King's College, Cambridge as mathematical scholar

1934 Graduates with distinction

1935 Is elected Fellow of King's College for dissertation on the Central Limit Theorem of Probability

1936 Goes to Princeton University where he works with Alonzo Church

1937 (January) His article "On Computable Numbers, with an Application to the Entscheidungsproblem" is published in *Proceedings of the London Mathematical Society*

Wins Procter Fellowship at Princeton

1938 Back in U.K. Attends course at the Government Code and Cypher School (G.C. & C.S.)

1939 Delivers undergraduate lecture-course in Cambridge and attends Wittgenstein's class on Foundations of Mathematics

4 September reports to G.C. & C.S. at Bletchley Park, in Buckinghamshire, where he heads work on German naval "Enigma" encoding machine

1942 Moves out of naval Enigma to become chief research consultant to G.C. & C.S.

In November sails to USA to establish liaison with American codebreakers

1943 January–March at Bell Laboratories in New York, working on speech-encypherment

1944 Seconded to the Special Communications Unit at Hanslope Park in north Buckinghamshire, where he works on his own speech-encypherment project *Delilah*

1945 With end of war is determined to design a prototype "universal machine" or "computer". In June is offered post with National Physical Laboratory at Teddington and begins work on ACE computer

1947 Severs relations with ACE project and returns to Cambridge

1948 Moves to Manchester University to work on prototype computer

1950 Publishes "Computing Machinery and Intelligence" in *Mind*

1951 Is elected FRS. Has become interested in problem of morphogenesis

1952 His article "The Chemical Basis of Morphogenesis" is published in *Philosophical Transactions of the Royal Society*

1954 Dies by his own hand in Wimslow (Cheshire) (7 June)

PREFACE TO THIS VOLUME

It may seem surprising that this collection of Alan Turing's work includes a whole volume devoted to biology, a subject in which he published only one paper. Biology was, however, far more important to Turing than is generally recognized. He had been interested in the subject right from his school days, and he had read, and been much impressed by, the book that has had such a strong influence on many theoretical biologists over the years, D'Arcy Thompson's (1917) classic *On Growth and Form*. He was also, like so many who work in biology, attracted by the sheer beauty of organisms. He wrote his (1952) paper not as a mathematical exercise, but because he saw the origin of biological form as one of the fundamental problems in science. And at the time of his death he was still working in biology, applying the theory he had derived to particular examples.

I found reading the archive material a fascinating experience. For while at first glance Turing's work on biology appears quite different from his other writings, it actually exhibits the features typical of all his work: his ability to identify a crucial problem in a field, his comparative lack of interest in what others were doing, his selection of an appropriate mathematical approach, and the great skill and evident ease with which he handled a wide range of mathematical techniques.

The biological work thus complements and completes the picture of Turing that the other volumes reveal: it shows the same style applied to a different problem. On the other hand, the nature of the material means that this volume differs from the others in two significant ways. Most of what the other three contain has appeared before; for the most part there seemed no reason to disagree with Turing's own judgement about what was worth publishing. The biological manuscripts, however, remained unpublished not by his choice but on account of his sudden death. I have therefore included a large amount of previously unpublished material. Much of it is from manuscripts prepared by N. Hoskin and B. Richards from a manuscript by Turing and from notes of his lectures, but some is by Turing himself. There is also a paper prepared by Richards from the work he did for his MSc. thesis under Turing's supervision but also not published. I have, however, omitted a number of fragments.

The manuscripts were never edited into a form ready for publication and so I have had to undertake this task myself. I have made some obvious minor corrections and filled in a few gaps where it was clear what was missing, but there are no significant alterations. My aim has been to produce as nearly as possible the papers that would have appeared had Turing lived. To avoid cluttering the text with indications of trivial deviations from the manuscript, I have not marked the corrections. Readers whose primary interest is historical are therefore warned that not only does the archive contain more material than is in this volume, but not everything that is here is word for word as it appears in the manuscripts.

In preparing this volume I have not felt the need to provide the sort of editorial notes that are found in the others. The mathematics is comparatively straightforward, and Turing was obviously trying to be as clear as he could for what he expected would be a mixed audience, very few of whom would know both mathematics and biology. Consequently, it is seldom necessary to explain what he is doing at any particular point. Instead, I have written introductions to the papers to put the work into context and to assist the reader with some points which are no longer as well known as they were when Turing was writing.

Acknowledgements

The Turing manuscripts are preserved in the library of King's College, Cambridge, and I am grateful to the College and the archivists for their cooperation. I am also grateful to the Royal Society of London, the Society for Experimental Biology, Bernard Richards and Alastair Wardlaw for agreeing to the publication of material in which they have interests.

Finally, I wish to thank Robin Gandy, Mae-Wan Ho, Bernard Richards and Alastair Wardlaw for helpful information and comments, and especially Nick Furbank, who organized the whole project and contributed so much to its success.

[X]

INTRODUCTION

Turing's work in biology illustrates just as clearly as his other work his ability to identify a fundamental problem and to approach it in a highly original way, drawing remarkably little from what others had done. He chose to work on the problem of form at a time when the majority of biologists were primarily interested in other questions. There are very few references in these papers, and most of them are for confirmation of details rather than for ideas which he was following up. In biology, as in almost everything else he did within science—or out of it—Turing was not content to accept a framework set up by others.

Even the fact that the mathematics in these papers is different from what he used in his other work is significant. For while it is not uncommon for a newcomer to make an important contribution to a subject, this is usually because he brings to it techniques and ideas which he has been using in his previous field but which are not known in the new one. Now much of Turing's career up to this point had been concerned with computers, from the hypothetical Turing machine to the real life Colossus, and this might have been expected to have led him to see the development of an organism from egg to adult as being programmed in the genes and to set out to study the structure of the programs. This would also have been in the spirit of the times, because the combining of Darwinian natural selection and Mendelian genetics into the synthetic theory of evolution had only been completed about ten years earlier, and it was in the very next year that Crick and Watson discovered the structure of DNA. Alternatively, Turing's experience in computing might have suggested to him something like what are now called cellular automata, models in which the fate of a cell is determined by the states of its neighbours through some simple algorithm, in a way that is very reminiscent of the Turing machine.

For Turing, however, the fundamental problem of biology had always been to account for pattern and form, and the dramatic progress that was being made at that time in genetics did not alter his view. And because he believed that the solution was to be found in physics and chemistry it was to these subjects and the sort of mathematics that could be applied to them that he turned. In my view, he was right, but even someone who disagrees must be impressed by the way in which he went directly to what *he* saw as the most important problem and set out to attack it with the tools that he judged appropriate to the task, rather than those which were easiest to hand or which others were already using. What is more, he understood the full significance of the problem in a way that many biologists did not and still do not. We can see this in the joint manuscript with Wardlaw which is included in this volume, but it is clear just from the comment he made to Robin Gandy (HODGES 1983, p. 431) that his new ideas were "intended to defeat the argument from design".

This single remark sums up one of the most crucial issues in contemporary biology. The argument from design was originally put forward as a scientific proof of the existence of God. The best known statement of it is William Paley's (1802) famous metaphor of the watchmaker. If we see a stone on some waste ground we do not wonder about it. If, on the other hand, we were to find a watch, with all its many parts combining so beautifully to achieve its purpose of keeping accurate time, we would be bound to infer that it had been designed and constructed by an intelligent being. Similarly, so the argument runs, when we look at an organism, and above all at a human being, how can we not believe that there must be an intelligent Creator?

Turing was not, of course, trying to refute Paley; that had been done almost a century earlier by Charles Darwin. But the argument from design had survived, and was, and indeed remains, still a potent force in biology. For the essence of Darwin's theory is that organisms are created by natural selection out of random variations. Almost any small variation can occur; whether it persists and so features in evolution depends on whether it is selected. Consequently we explain how a certain feature has evolved by saying what advantage it gives to the organism, i.e. what purpose it serves, just as if we were explaining why the Creator has designed the organism in that way. Natural selection thus takes over the role of the Creator, and becomes "The Blind Watchmaker" (DAWKINS 1986).

Not all biologists, however, have accepted this view. One of the strongest dissenters was D'Arcy Thompson (1917), who insisted that biological form is to be explained chiefly in the same way as inorganic form, i.e., as the result of physical and chemical processes. The primary task of the biologist is to discover the set of forms that are likely to appear. Only then is it worth asking which of them will be selected. Turing, who had been very much influenced by D'Arcy Thompson, set out to put the program into practice. Instead of asking why a certain arrangement of leaves is especially advantageous to a plant, he tried to show that it was a natural consequence of the process by which the leaves are produced. He did not in fact achieve his immediate aim, and indeed more than thirty-five years later the problem of phyllotaxis has still not been solved. On the other hand, the reaction-diffusion model has been applied to many other problems of pattern and form and Turing structures (as they are now called) have been observed experimentally (CASTETS et al. 1990), so Turing's idea has been vindicated.

Outline of the Volume

The first paper in this volume, *The Chemical Basis of Morphogenesis*, is the only one that Turing ever published in biology. It sets out the reaction-diffusion theory of pattern formation and gives an example of its application. It is followed by a paper which was intended to be published jointly by Turing and the botanist C.W. Wardlaw. This gives more of the biological background and also a brief non-technical account of the theory itself. The version printed here is a draft; Wardlaw later published the work in a slightly modified form (WARDLAW, 1953). After this comes Turing's major unfinished work, *Morphogen Theory of Phyllotaxis*. This is in three parts, the first a geometrical description of the patterns to be explained and the second an application of the reaction-diffusion theory to the problem. It was left in typescript form, mostly prepared by N. Hoskin from notes and lectures by Turing. There are two versions of part II in the archive, one a draft of the other. The revised version ends after §3, and so I have used the draft for the later sections.

The third part is not by Turing himself but by his student Bernard Richards, now Professor of Computing at the University of Manchester Institute of Science and Technology. The problem was, however, suggested by Turing and the biological content is due to him: Richards carried out the mathematical and numerical work as the project for his MSc. thesis. Finally there is the incomplete *Outline of the Development of the Daisy*. As I explained in the preface, I am not including everything in the Archive, but I felt that this material does stand on its own. It gives us an idea of how Turing meant to proceed and it also reminds us that Turing was interested not just in mathematics but also in real flowers, an interest that goes back to his childhood if we may judge by the sketch that serves as the frontispiece.

The Chemical Basis of Morphogenesis

The development of any organism, and above all a complex one such as a human being, is a truly remarkable process. We each begin as a single cell and eventually become an adult made up of approximately 10^{15} cells of about 200 different types organized in a very complicated arrangement and able to cooperate to carry out many vital functions. This would be an impressive enough accomplishment if it were done under the supervision of an intelligent craftsman; in fact it happens through nothing more than a series of interrelated physical and chemical processes. The genes play an important role in this, but we cannot just say that the genes create the form and let it go at that. The genes can only influence development through their effects on chemical reactions, and they themselves have to be turned on and off at appropriate times. Important though developmental genetics is, ultimately it is the physics and chemistry that we have to understand.

While the later stages of development are often complicated and hard to understand in detail, perhaps the greatest difficulty in principle is at the very beginning. Once a pattern of some sort has been established, it can serve as the basis for the next stage, and so on. But how does the process start? The original cell is not, to be sure, totally symmetric, it has a polarity induced by the point of entry of the sperm, but this does not seem enough to determine the structure that is to appear. How does a pattern appear in a region which has nothing to serve as a template—or, equivalently, where does the template come from? This was what Turing saw as the fundamental problem.

He found an answer in the bifurcation properties of the solutions of differential equations. Applied mathematicians had been aware for many years that when a parameter of a system passes through a certain critical value there can be a qualitative change in behaviour as a previously stable state becomes unstable. The archetypal example, first studied by Euler more than two centuries earlier, is the sudden buckling of a beam when it is overloaded.

Turing used the idea in a different way. He constructed a system of simple partial differential equations that can plausibly be supposed to govern the concentration of some chemical substance, C say, throughout a region. The equations were designed in such a way that $C =$ constant is always a solution, but it is not always stable. Then simply by varying the value of a parameter (which in a real situation could correspond to the rate of some reaction or the size of the region) one obtains either a homogeneous distribution or else a definite and predictable pattern.

That Turing used a chemical model should not be taken to imply that he saw morphogenesis * as a purely chemical phenomenon. On the contrary

---

* Strictly speaking, morphogenesis is the generation of *form*, which is not actually what this paper is about. Turing was aware of this, and explained in *A Diffusion Reaction Theory of Morphogenesis in Plants*, this volume, footnote on p. 38, why he still preferred to use the term, but modern workers generally write of "pattern formation"—though they do follow Turing in referring to the crucial chemical substances as morphogens.

he explicitly acknowledged the importance of mechanical forces and also of the electrical properties of cells. He confined himself to the chemical aspects because he recognized that he could only hope to make progress in simplified cases, and chemistry seemed the appropriate place to begin since the action of the genes would presumably be chemical. All this is carefully explained in the first section of the paper. Here we see an important characteristic of Turing's work in biology: he was willing to make simplifying assumptions where they were necessary to allow progress, but he was always careful both to point out the assumptions and to provide some justification for them.

Turing was clearly determined to make his work comprehensible to as many readers as possible. So the second and third sections are mathematical and chemical instruction for the reader with insufficient knowledge of one or the other, while section four is a non-mathematical description of the basic idea of the paper, including an ingenious account in lay terms of the idea of a bifurcation parameter. Even then, Turing was not ready to embark on the major part of his work until he had dealt with a possible objection, viz. that the model he was about to propose can account for a loss of symmetry but not for the systematic bias towards either left- or right-handedness which is so common in organisms. Whether or not Turing's explanation turns out to be correct, it is typical that he should have seen the difficulty, acknowledged that his theory cannot deal with it, and proposed other effects which might overcome it, before going on to develop his model.

Finally, Turing introduced the model itself, considering first a ring of N cells and then a continuous ring of tissue. In both cases he solved the equations for small perturbations about the uniform equilibrium solution and he found that they were quite similar. As he pointed out, this is not surprising, since the latter situation is a limiting case of the former, but it has a significance that he did not mention. Pattern formation often occurs in tissues which are not divided into cells. This is hard to explain by models which are based on differential gene expression and so it an argument in favour of an approach which, like Turing's, does not depend on the existence of separate cells. Conversely, that the mechanism can produce patterns whether cells exist or not suggests that cells and other divisions may be less important in development than is generally thought.

The paper then becomes more technical, with more mathematical results, a discussion of the problem of the effects of disturbances, and the results of some numerical calculations, including a figure showing a dappled pattern in two dimensions "obtained in a few hours by a manual [!] computation". In his summary of this part of the paper, Turing defended the linearity assumption, on which almost everything depends, on the grounds that "the patterns produced in the early stages when it is valid may be expected to have strong qualitative similarity to those prevailing in the later stages when it does not". He gave no particular evidence for this idea, but it appears reasonable and is probably largely true; MURRAY (1981) has compared the properties of some linearized solutions with solutions of the full equations obtained numerically and much of the qualitative behaviour is indeed the same. Turing considered the passage from stability to instability in a single cell as the equilibrium concentrations and reaction rates vary as "the least interesting of the cases" but he did remark that it could produce dappling and also that the patterns would have to be laid down when the foetus was small enough that the morphogen could diffuse across it in a relatively short time. Both these comments have been borne out by later work.

Next, there is a discussion of an application of the work to real problems involving rings of cells, or something close to them, like the tentacles of Hydra and the whorls of leaves of plants such as Woodruff, and finally to gastrulation.

In the last section, Turing acknowledged that his approach is, on account of the complexity of the calculations, unlikely to lead to any theory of pattern formation, but only to results in particular cases. He took the view that this was not such a great disadvantage, since the computations would probably be illuminating enough. And indeed, one would expect that a number of well chosen examples would give a reasonable understanding of the ways in which processes of this kind work, and the sorts of patterns they can and cannot produce. It may be that mathematical tools that have been developed since this paper was written, and others that are yet to come, will eventually provide the rigorous results that Turing did not expect to see. We may yet hope for definitive lists of patterns along the lines of THOM's (1972) classification of the elementary catastrophes. But whether or not this happens, *The Chemical Basis of Morphogenesis* is a classic paper. It is still very frequently cited (more than the rest of Turing's works taken together, though I would not claim that as evidence of relative importance) and the reaction-diffusion mechanism, or "Turing-type" model as it is often called, has become one of the standard models of theoretical biology.

The paper is not, however, only about one particular model. It contains two basic ideas, of which the proposed equations are really just examples. First, a pattern can appear through an instability of the constant solution of perfectly simple and plausible differential equations. There is no need to postulate something outlandish. Second, the pattern is then determined by the equations and the shape of the region. It does not have to be specified by some other process, such as the switching on of different genes in different cells; as Turing showed, the model works whether the region is divided into cells or not. Whatever the fate of the reaction-diffusion model itself, these principles are certain to remain fundamental in biological modelling.

Morphogen Theory of Phyllotaxis

It is not hard to imagine why Turing chose the arrangement of leaves on plants as the first application of his theory. Phyllotaxis is a classical problem which remains unsolved to this day, despite the efforts of many workers. Yet at the same time it is hard to believe that it does not have a straightforward solution, if only one were clever enough to find it. The phenomenon to be explained is the occurrence of a small number of regular patterns on a simply shaped and accessible surface. The pattern on a mature specimen is essentially that which is laid down in the first place, which is not so in many other developmental processes. And for a mathematician there is the additional twist that the Fibonacci sequence is involved.

Turing's attempt on the problem consists of two parts. The first is a detailed geometrical analysis of the patterns, and the second is the beginning of an application of the (1952) theory to explain them. While the latter, though incomplete, is quite straightforward and self-contained, the former requires some further explanation. Many readers may not know much about phyllotaxis, and most of those that do will probably be accustomed to accounts written by botanists, who usually approach the problem slightly differently. In particular, because Turing was setting out to investigate as deeply as possible the patterns he was hoping to explain, he chose to represent the leaves as the points of a lattice. This amounts to considering the mature stem as a cylinder, unrolling the surface onto the plane, and then repeating the pattern infinitely many times. There are obvious mathematical advantages in this, and Turing is not the only author to have done it, but it does mean that there are some differences between his approach and the usual botanists' picture, which is based on a cross section. Above all, the connection with the Fibonacci sequence is far less obvious. To assist the reader, therefore, I provide below an outline introduction to phyllotaxis with definitions of the terms that are used, referring where necessary to both representations. I also include a brief account of continued fractions and an explanation of how the Fibonacci numbers enter into the problem and how they are connected with the Fibonacci angle.

Phyllotaxis: There are several common forms of phyllotaxis. In some plants, such as grasses and peas, each leaf is at an angle of 180° from the one before it on the stem. This is called distichous phyllotaxis. In another form, known as decussate phyllotaxis and found in, for example, trees like the ash and horse chestnut, the leaves occur in opposing pairs, with each pair in a plane at right angles to the one before. In most flowering plants, however, and in conifers and various other families, the leaves are arranged around the stem in such a way that it is possible to draw a single spiral, called the *fundamental* (also *genetic, ontogenetic* or *generative*) spiral which passes through the centres of all of them in the order in which they appeared. Since the time interval between successive appearances of primordia (the *plastochrone*) is approximately constant, so too are the distances and the angles between them.

The angle between successive primordia, or leaf centres in the mature plant, is called the *divergence angle*. To specify the arrangement completely a second coordinate is required and if we are studying the cross section a convenient choice is the *plastochrone ratio* (RICHARDS 1984), the ratio of the transverse distances from the centre of successive primordia. In a uniform system it is a measure of the radial expansion of the apex during one plastochrone. Because he was concerned with the side view, Turing

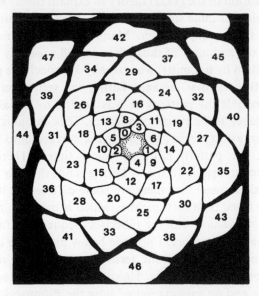

Fig. 1. Transection of the apical bud of a young seedling of *Prinus pinea*. The leaves are numbered in order of formation. The contact parastichy numbers are 5 and 8. Redrawn after CHURCH (1920).

[XVIII]

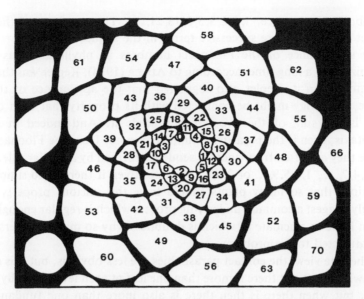

Fig. 2. Transection of the apical bud of a young seedling of *Araucaria excelsa*. The leaves are numbered in order of formation. The contact parastichy numbers are 7 and 11. Redrawn after CHURCH (1920).

(§2) used instead what he called the *plastochrone distance*, which is measured along the surface of the stem; in the case of a cylindrical stem it is along one of the generators.

On account of the regularity, many spirals other than the fundamental one can be drawn through primordia; any such spiral is called a *parastichy*. Two examples are shown in Figs. 1 and 2. These are illustrations of the kind commonly found in biological works and show a transection of the apical bud. Drawn in this way, one's eye is immediately caught not by the fundamental spiral but by the two parastichies, one spiralling clockwise and one anticlockwise, that pass through primordia that are actually in contact and are therefore called *contact parastichies*. In the mature plant they are the parastichies which pass through a given leaf and one of the two or three adjacent leaves above or below it, and they are then sometimes referred to as the *conspicuous opposed parastichy pair*, because the leaves are not actually in contact as the primordia were. If the primordia are numbered in order of formation, or in the case of a mature plant if the leaves are numbered in order along the stem, it is easily seen that the difference in number between successive primordia on a parastichy will be constant. What is surprising is that in the vast majority of cases, these numbers, called the numbers or orders of the parastichies, are members of the Fibonacci sequence, $1, 1, 2, 3, 5, 8, 13, 21, \ldots$. The numbers of the two contact parastichies are successive terms in the sequence.

The significance of the Fibonacci numbers in phyllotaxis has been recognized for a long time; according to ADLER (1974), Kepler was the first to comment on it. Kepler also suggested that the appearance of the sequence in biology might be connected with its property that each of the terms is the sum of the two which precede it. And indeed in spiral phyllotaxis, even if the parastichy numbers are not from the Fibonacci sequence they are often from another sequence formed by a similar rule, such as $1, 3, 4, 7, \ldots$, $1, 4, 5, 9, \ldots$ or $2, 5, 7, 12, \ldots$ etc. Even if there is no simple explanation, that so many plants should have this curious property does strongly suggest a common underlying process which is regular enough that we can hope to elucidate it, which is doubtless why so many workers have been attracted to the problem.

In the side view, the contact parastichies are less obvious, but it is easier to see whether or not there is more than one leaf at each level. Usually there is not, but when there is then there is also more than one fundamental spiral. Turing denoted the number of leaves at each level by J, and called it the *jugacy*, because the cases $J = 2$, $J = 3$, $J > 3$ are commonly referred to as bijugate, trijugate and multijugate, respectively. In Fig. 3 we have supposed that $J = 1$. There is consequently only one fundamental spiral, and we take this to be a helix with the leaves at equal intervals along it so that the leaves form a cylindrical lattice. Figure 3(b) shows the equivalent plane lattice.

A parastichy will not in general pass through a leaf at every level. If it

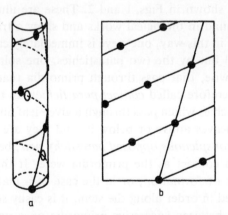

Fig. 3. (a) Side view of an idealized stem and (b) the equivalent plane lattice. The contact parastichy numbers are 1 and 2. Both the generative spiral and a parastichy of order 3 are shown.

passes through a leaf at every nth level only, then it is called a parastichy of order n. Turing called n the *parastichy number*. There are altogether nJ parastichies of order n in a phyllotaxis of jugacy J. They are all parallel, and every leaf lies on exactly one of them, so they partition the set of leaves. A collection of mJ m-order parastichies and nJ n-order parastichies with m and n chosen so that one set is clockwise and one anticlockwise is called an *opposed parastichy pair* of order (m, n). There need not be a leaf at every intersection of the two spirals of such a pair, but if there is, the pair is called *visible*. In the lattice representation the contact parastichies are defined as the parastichy pair defined by a leaf together with its nearest neighbours to the right and to the left. This is of course equivalent to the definition given above.

The Fibonacci sequence: Almost everyone who writes about phyllotaxis points out the striking property that the contact parastichies typically have numbers which are members of the Fibonacci sequence and that the divergence angle (i.e. the angle between successive primordia) is then close to the Fibonacci angle, approximately $137.51°$. They also generally mention that other divergence angles occur, though less frequently, that these are approximately $99.50°$ and $77.96°$, and that the contact parastichies then have numbers from the series $1, 3, 4, 7, 11, \ldots$ (the anomalous or first accessory series) or $1, 4, 5, 9, 14, \ldots$, respectively. The latter two series satisfy the same recurrence relation as the Fibonacci series, viz. $u_n = u_{n-1} + u_{n-2}$.

These statements are usually made without any explanation. At the time when Turing was writing, it may have been safe to suppose that most readers, at least those who were mathematicians, would be familiar with the connection between the Fibonacci sequence and the particular angle. Because this is less true today, and even continued fractions (which are used in §10) are no longer a standard part of mathematics syllabi, we give a brief account here.

A continued fraction is a fraction of the form

$$a_0 + \cfrac{1}{a_1 + \cfrac{1}{a_2 + \cfrac{1}{a_3 + \cfrac{1}{a_4 + \cdots}}}}$$

where $a_0, a_1, a_2, \ldots, a_n, \ldots$ are real numbers all of which, with the possible exception of a_0, are positive. Because this is an awkward expression to write or set in type it is usual to employ a conventional notation, such as the one Turing used in §10 or the even simpler $[a_0; a_1, a_2, ...]$. The numbers a_n are called the partial quotients of the fraction. If there are only a finite number of non-zero partial quotients the continued fraction is said to be finite; if all the partial quotients are integers it is said to be simple. The finite continued fraction obtained from an infinite one by cutting off the expansion after the nth partial quotient, a_n, is called the nth convergent of the continued fraction.

It is not difficult to show (see, e.g. BURTON (1976), p. 306) that the nth convergent of the simple continued fraction $[a_0; a_1, a_2, ...]$ is given by

$$C_n = p_n/q_n$$

where

$$p_0 = a_0, \qquad q_0 = 1,$$
$$p_1 = a_1 a_0 + 1, \qquad q_1 = a_1,$$
$$p_k = a_k p_{k-1} + p_{k-2}, \qquad q_k = a_k q_{k-1} + q_{k-2}, \quad k > 1.$$

It can also be shown that p_k and q_k are relatively prime, so the formula gives the convergents in their lowest terms. If $a_0 = 0$ and $a_k = 1$ for all $k > 0$, the recurrence relations for p_k and q_k generate Fibonacci series, and the successive convergents are

$$C_n = u_{n+1}/u_n$$

where u_n denotes the nth Fibonacci number.

Every rational number can be written as a finite simple continued fraction in two closely related ways. Every irrational number has a unique representation as an infinite simple continued fraction. Much of the importance of continued fractions arises from the fact that if an irrational number x is expressed as a continuous fraction, the convergents p_n/q_n are the best approximations to x in the sense that each of them gives the closest approximation to x among all rational numbers with denominators q_n or less. This property makes continued fractions useful in numerical analysis, and it is also the reason that they are connected with phyllotaxis.

The contact parastichies are the parastichies formed by adjacent leaf bases. Now which bases are adjacent to a given base O depends not only on the angular separation between them but also on the pitch of the helix (see Fig. 4). On the other hand, it is clear that a leaf base is a candidate for being adjacent only if it is closer to the generator (i.e. the vertical line)

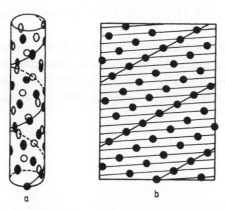

Fig. 4. (a) Side view of an idealized stem and (b) the equivalent plane lattice. As Fig. 3 except for the pitch of the helix. The contact parastichy numbers are 3 and 5.

through O than is any previous base on the generative spiral, or at least any previous base on the same side of the generator.

Now the leaves are generally equally spaced along the generative spiral. Let their angular separation, i.e. the divergence angle, be θ, and let $\alpha = \theta/2\pi$. In finding leaves that are close to the original vertical line we are looking for integers a, b such that $a/b \doteqdot 2\alpha\pi$. Here b is the number of the leaf in sequence along the generative spiral and a is the number of rotations the spiral has made around the stem. The closest rational approximations to α are the convergents of its expansion as a continued fraction. Given any α, therefore, we can immediately calculate the convergents, and because the convergents are automatically in their lowest terms this gives us p_k and q_k separately. The sequence p_k gives the sequence of leaves successively closest to the vertical line, i.e. the set of possible contact parastichy numbers for the given divergence angle.

Conversely, if we are given the complete set of possible contact parastichy numbers we can work out the divergence angle by solving the recurrence relation for the a_k. In particular, if $p_k = u_k$, the kth Fibonacci number, then the continued fraction is $[0, 1, 1, 1, ...]$, which can be shown to be equal to $(1 + \sqrt{5})/2$, or about 1.618, the so-called "golden mean". This implies a divergence angle of about 582.5°, or, equivalently, 137.5°. For this divergence angle, therefore, the contact parastichy numbers will always be Fibonacci numbers, though which ones they will be will depend on the pitch of the helix. The angle 137.5° is called the Fibonacci angle.

The next most common spirals have parastichy numbers from the subsidiary series $1, 3, 4, 7, 11, ...$. This corresponds to the continued fraction $[0, 3, 1, 1, 1, ...]$ which implies a divergence angle of approximately 99.5°.

Also observed is $1,4,5,9,14,\ldots$, with the continued fraction $[0,4,1,1,1,\ldots]$ and angle $77.96°$.

For more on continued fractions and the related sequences, see BURTON (1976) or almost any other elementary book on number theory. The "simple inductive argument" referred to in the text (p. 66) and proofs of the results mentioned above can be found in Burton's book.

CONTENTS

CONTENTS

THE CHEMICAL BASIS OF MORPHOGENESIS

By A. M. TURING, F.R.S. *University of Manchester*

(*Received* 9 *November* 1951—*Revised* 15 *March* 1952)

It is suggested that a system of chemical substances, called morphogens, reacting together and diffusing through a tissue, is adequate to account for the main phenomena of morphogenesis. Such a system, although it may originally be quite homogeneous, may later develop a pattern or structure due to an instability of the homogeneous equilibrium, which is triggered off by random disturbances. Such reaction-diffusion systems are considered in some detail in the case of an isolated ring of cells, a mathematically convenient, though biologically unusual system. The investigation is chiefly concerned with the onset of instability. It is found that there are six essentially different forms which this may take. In the most interesting form stationary waves appear on the ring. It is suggested that this might account, for instance, for the tentacle patterns on *Hydra* and for whorled leaves. A system of reactions and diffusion on a sphere is also considered. Such a system appears to account for gastrulation. Another reaction system in two dimensions gives rise to patterns reminiscent of dappling. It is also suggested that stationary waves in two dimensions could account for the phenomena of phyllotaxis.

The purpose of this paper is to discuss a possible mechanism by which the genes of a zygote may determine the anatomical structure of the resulting organism. The theory does not make any new hypotheses; it merely suggests that certain well-known physical laws are sufficient to account for many of the facts. The full understanding of the paper requires a good knowledge of mathematics, some biology, and some elementary chemistry. Since readers cannot be expected to be experts in all of these subjects, a number of elementary facts are explained, which can be found in text-books, but whose omission would make the paper difficult reading.

1. A MODEL OF THE EMBRYO. MORPHOGENS

In this section a mathematical model of the growing embryo will be described. This model will be a simplification and an idealization, and consequently a falsification. It is to be hoped that the features retained for discussion are those of greatest importance in the present state of knowledge.

The model takes two slightly different forms. In one of them the cell theory is recognized but the cells are idealized into geometrical points. In the other the matter of the organism is imagined as continuously distributed. The cells are not, however, completely ignored, for various physical and physico-chemical characteristics of the matter as a whole are assumed to have values appropriate to the cellular matter.

With either of the models one proceeds as with a physical theory and defines an entity called 'the state of the system'. One then describes how that state is to be determined from the state at a moment very shortly before. With either model the description of the state consists of two parts, the mechanical and the chemical. The mechanical part of the state describes the positions, masses, velocities and elastic properties of the cells, and the forces between them. In the continuous form of the theory essentially the same information is given in the form of the stress, velocity, density and elasticity of the matter. The chemical part of the state is given (in the cell form of theory) as the chemical composition of each separate cell; the diffusibility of each substance between each two adjacent cells must also

[[1]]

be given. In the continuous form of the theory the concentrations and diffusibilities of each substance have to be given at each point. In determining the changes of state one should take into account

(i) The changes of position and velocity as given by Newton's laws of motion.

(ii) The stresses as given by the elasticities and motions, also taking into account the osmotic pressures as given from the chemical data.

(iii) The chemical reactions.

(iv) The diffusion of the chemical substances. The region in which this diffusion is possible is given from the mechanical data.

This account of the problem omits many features, e.g. electrical properties and the internal structure of the cell. But even so it is a problem of formidable mathematical complexity. One cannot at present hope to make any progress with the understanding of such systems except in very simplified cases. The interdependence of the chemical and mechanical data adds enormously to the difficulty, and attention will therefore be confined, so far as is possible, to cases where these can be separated. The mathematics of elastic solids is a well-developed subject, and has often been applied to biological systems. In this paper it is proposed to give attention rather to cases where the mechanical aspect can be ignored and the chemical aspect is the most significant. These cases promise greater interest, for the characteristic action of the genes themselves is presumably chemical. The systems actually to be considered consist therefore of masses of tissues which are not growing, but within which certain substances are reacting chemically, and through which they are diffusing. These substances will be called morphogens, the word being intended to convey the idea of a form producer. It is not intended to have any very exact meaning, but is simply the kind of substance concerned in this theory. The evocators of Waddington provide a good example of morphogens (Waddington 1940). These evocators diffusing into a tissue somehow persuade it to develop along different lines from those which would have been followed in its absence. The genes themselves may also be considered to be morphogens. But they certainly form rather a special class. They are quite indiffusible. Moreover, it is only by courtesy that genes can be regarded as separate molecules. It would be more accurate (at any rate at mitosis) to regard them as radicals of the giant molecules known as chromosomes. But presumably these radicals act almost independently, so that it is unlikely that serious errors will arise through regarding the genes as molecules. Hormones may also be regarded as quite typical morphogens. Skin pigments may be regarded as morphogens if desired. But those whose action is to be considered here do not come squarely within any of these categories.

The function of genes is presumed to be purely catalytic. They catalyze the production of other morphogens, which in turn may only be catalysts. Eventually, presumably, the chain leads to some morphogens whose duties are not purely catalytic. For instance, a substance might break down into a number of smaller molecules, thereby increasing the osmotic pressure in a cell and promoting its growth. The genes might thus be said to influence the anatomical form of the organism by determining the rates of those reactions which they catalyze. If the rates are assumed to be those determined by the genes, and if a comparison of organisms is not in question, the genes themselves may be eliminated from the discussion. Likewise any other catalysts obtained secondarily through the agency of the genes may equally be ignored, if there is no question of their concentrations varying. There may, however, be some other morphogens, of the nature of evocators, which cannot be altogether forgotten, but whose role may nevertheless be subsidiary, from the point of view of the formation of a particular organ. Suppose, for instance, that a 'leg-evocator' morphogen were being produced in a certain region of an embryo, or perhaps diffusing into it, and that an attempt was being made to explain the mechanism by which the leg was formed in the presence of the evocator. It would then be reasonable to take the distribution of the evocator in space and time as given in advance and to consider the chemical reactions set in train by it. That at any rate is the procedure adopted in the few examples considered here.

2. MATHEMATICAL BACKGROUND REQUIRED

The greater part of this present paper requires only a very moderate knowledge of mathematics. What is chiefly required is an understanding of the solution of linear differential equations with constant coefficients. (This is also what is chiefly required for an understanding of mechanical and electrical oscillations.) The solution of such an equation takes the form of a sum $\Sigma A\, e^{bt}$, where the quantities A, b may be complex, i.e. of the form $\alpha + i\beta$, where α and β are ordinary (real) numbers and $i = \sqrt{-1}$. It is of great importance that the physical significance of the various possible solutions of this kind should be appreciated, for instance, that

(a) Since the solutions will normally be real one can also write them in the form $\mathcal{R}\Sigma A\, e^{bt}$ or $\Sigma \mathcal{R} A\, e^{bt}$ ($\mathcal{R}$ means 'real part of').

(b) That if $A = A'\, e^{i\phi}$ and $b = \alpha + i\beta$, where A', α, β, ϕ are real, then

$$\mathcal{R} A\, e^{bt} = A'\, e^{\alpha t} \cos(\beta t + \phi).$$

Thus each such term represents a sinusoidal oscillation if $\alpha = 0$, a damped oscillation if $\alpha < 0$, and an oscillation of ever-increasing amplitude if $\alpha > 0$.

(c) If any one of the numbers b has a positive real part the system in question is unstable.

(d) After a sufficiently great lapse of time all the terms $A\, e^{bt}$ will be negligible in comparison with those for which b has the greatest real part, but unless this greatest real part is itself zero these dominant terms will eventually either tend to zero or to infinite values.

(e) That the indefinite growth mentioned in (b) and (d) will in any physical or biological situation eventually be arrested due to a breakdown of the assumptions under which the solution was valid. Thus, for example, the growth of a colony of bacteria will normally be taken to satisfy the equation $dy/dt = \alpha y$ ($\alpha > 0$), y being the number of organisms at time t, and this has the solution $y = A\, e^{\alpha t}$. When, however, the factor $e^{\alpha t}$ has reached some billions the food supply can no longer be regarded as unlimited and the equation $dy/dt = \alpha y$ will no longer apply.

The following relatively elementary result will be needed, but may not be known to all readers:

$$\sum_{r=1}^{N} \exp\left[\frac{2\pi i r s}{N}\right] = 0 \quad \text{if} \quad 0 < s < N,$$

but
$$= N \quad \text{if} \quad s = 0 \text{ or } s = N.$$

The first case can easily be proved when it is noticed that the left-hand side is a geometric progression. In the second case all the terms are equal to 1.

The relative degrees of difficulty of the various sections are believed to be as follows. Those who are unable to follow the points made in this section should only attempt §§ 3, 4, 11, 12, 14 and part of § 13. Those who can just understand this section should profit also from §§ 7, 8, 9. The remainder, §§ 5, 10, 13, will probably only be understood by those definitely trained as mathematicians.

3. CHEMICAL REACTIONS

It has been explained in a preceding section that the system to be considered consists of a number of chemical substances (morphogens) diffusing through a mass of tissue of given geometrical form and reacting together within it. What laws are to control the development of this situation? They are quite simple. The diffusion follows the ordinary laws of diffusion, i.e. each morphogen moves from regions of greater to regions of less concentration, at a rate proportional to the gradient of the concentration, and also proportional to the 'diffusibility' of the substance. This is very like the conduction of heat, diffusibility taking the place of conductivity. If it were not for the walls of the cells the diffusibilities would be inversely proportional to the square roots of the molecular weights. The pores of the cell walls put a further handicap on the movement of the larger molecules in addition to that imposed by their inertia, and most of them are not able to pass through the walls at all.

The reaction rates will be assumed to obey the 'law of mass action'. This states that the rate at which a reaction takes place is proportional to the concentrations of the reacting substances. Thus, for instance, the rate at which silver chloride will be formed and precipitated from a solution of silver nitrate and sodium chloride by the reaction

$$Ag^+ + Cl^- \to AgCl$$

will be proportional to the product of the concentrations of the silver ion Ag^+ and the chloride ion Cl^-. It should be noticed that the equation

$$AgNO_3 + NaCl \to AgCl + NaNO_3$$

is not used because it does not correspond to an actual reaction but to the final outcome of a number of reactions. The law of mass action must only be applied to the *actual* reactions. Very often certain substances appear in the individual reactions of a group, but not in the final outcome. For instance, a reaction $A \to B$ may really take the form of two steps $A + G \to C$ and $C \to B + G$. In such a case the substance G is described as a catalyst, and as catalyzing the reaction $A \to B$. (Catalysis according to this plan has been considered in detail by Michaelis & Menten (1913).) The effect of the genes is presumably achieved almost entirely by catalysis. They are certainly not permanently used up in the reactions.

Sometimes one can regard the effect of a catalyst as merely altering a reaction rate. Consider, for example, the case mentioned above, but suppose also that A can become detached from G, i.e. that the reaction $C \to A + G$ is taken into account. Also suppose that the reactions $A + G \rightleftharpoons C$ both proceed much faster than $C \to B + G$. Then the concentrations of A, G, C will be related by the condition that there is equilibrium between the reactions $A + G \to C$ and $C \to A + G$, so that (denoting concentrations by square brackets) $[A][G] = k[C]$ for some constant k. The reaction $C \to B + G$ will of course proceed at a rate proportional to $[C]$, i.e. to $[A][G]$. If the amount of C is always small compared with the amount of G one can say that the presence of the catalyst and its amount merely alter the mass action constant for the reaction $A \to B$, for the whole proceeds at a rate proportional to $[A]$. This situation does not, however, hold invariably. It may well happen that nearly all of G takes the combined form C so long as any of A is left. In this case the reaction proceeds at a rate independent of the concentration of A until A is entirely consumed. In either of these cases the rate of the complete group of reactions depends only on the concentrations of the reagents, although usually not according to the law of mass action applied crudely to the chemical equation for the whole group. The same applies in any case where all reactions of the group with one exception proceed at speeds much greater than that of the exceptional one. In these cases the rate of the reaction is a function of the concentrations of the reagents. More generally again, no such approximation is applicable. One simply has to take all the actual reactions into account.

According to the cell model then, the number and positions of the cells are given in advance, and so are the rates at which the various morphogens diffuse between the cells. Suppose that there are N cells and M morphogens. The state of the whole system is then given by MN numbers, the quantities of the M morphogens in each of N cells. These numbers change with time, partly because of the reactions, partly because of the diffusion. To determine the part of the rate of change of one of these numbers due to diffusion, at any one moment, one only needs to know the amounts of the same morphogen in the cell and its neighbours, and the diffusion coefficient for that morphogen. To find the rate of change due to chemical reaction one only needs to know the concentrations of all morphogens at that moment in the one cell concerned.

This description of the system in terms of the concentrations in the various cells is, of course, only an approximation. It would be justified if, for instance, the contents were perfectly stirred. Alternatively, it may often be justified on the understanding that the 'concentration in the cell' is the concentration at a certain representative point, although the idea of 'concentration at a point' clearly itself raises difficulties. The author believes that the approximation is a good one, whatever argument is used to justify it, and it is certainly a convenient one.

It would be possible to extend much of the theory to the case of organisms immersed in a fluid, considering the diffusion within the fluid as well as from cell to cell. Such problems are not, however, considered here.

4. THE BREAKDOWN OF SYMMETRY AND HOMOGENEITY

There appears superficially to be a difficulty confronting this theory of morphogenesis, or, indeed, almost any other theory of it. An embryo in its spherical blastula stage has spherical symmetry, or if there are any deviations from perfect symmetry, they cannot be regarded as of any particular importance, for the deviations vary greatly from embryo to embryo within a species, though the organisms developed from them are barely distinguishable. One may take it therefore that there is perfect spherical symmetry. But a system which has spherical symmetry, and whose state is changing because of chemical reactions and diffusion, will remain spherically symmetrical for ever. (The same would hold true if the state were changing according to the laws of electricity and magnetism, or of quantum mechanics.) It certainly cannot result in an organism such as a horse, which is not spherically symmetrical.

There is a fallacy in this argument. It was assumed that the deviations from spherical symmetry in the blastula could be ignored because it makes no particular difference what form of asymmetry there is. It is, however, important that there are *some* deviations, for the system may reach a state of instability in which these irregularities, or certain components of them, tend to grow. If this happens a new and stable equilibrium is usually reached, with the symmetry entirely gone. The variety of such new equilibria will normally not be so great as the variety of irregularities giving rise to them. In the case, for instance, of the gastrulating sphere, discussed at the end of this paper, the direction of the axis of the gastrula can vary, but nothing else.

The situation is very similar to that which arises in connexion with electrical oscillators. It is usually easy to understand how an oscillator keeps going when once it has started, but on a first acquaintance it is not obvious how the oscillation begins. The explanation is that there are random disturbances always present in the circuit. Any disturbance whose frequency is the natural frequency of the oscillator will tend to set it going. The ultimate fate of the system will be a state of oscillation at its appropriate frequency, and with an amplitude (and a wave form) which are also determined by the circuit. The phase of the oscillation alone is determined by the disturbance.

If chemical reactions and diffusion are the only forms of physical change which are taken into account the argument above can take a slightly different form. For if the system originally has no sort of geometrical symmetry but is a perfectly homogeneous and possibly irregularly shaped mass of tissue, it will continue indefinitely to be homogeneous. In practice, however, the presence of irregularities, including statistical fluctuations in the numbers of molecules undergoing the various reactions, will, if the system has an appropriate kind of instability, result in this homogeneity disappearing.

This breakdown of symmetry or homogeneity may be illustrated by the case of a pair of cells originally having the same, or very nearly the same, contents. The system is homogeneous: it is also symmetrical with respect to the operation of interchanging the cells. The contents of either cell will be supposed describable by giving the concentrations X and Y of two morphogens. The chemical reactions will be supposed such that, on balance, the first morphogen (X) is produced at the rate $5X - 6Y + 1$ and the second (Y) at the rate $6X - 7Y + 1$. When, however, the strict application of these formulae would involve the concentration of a morphogen in a cell becoming negative, it is understood that it is instead destroyed only at the rate at which it is reaching that cell by diffusion. The first morphogen will be supposed to duffuse at the rate $0 \cdot 5$ for unit difference of concentration between the cells, the second, for the same difference, at the rate $4 \cdot 5$. Now if both morphogens have unit concentration in both cells there is equilibrium. There is no resultant passage of either morphogen across the cell walls, since there is no concentration difference, and there is no resultant production (or destruction) of either morphogen in either cell since $5X - 6Y + 1$ and $6X - 7Y + 1$ both have the value zero for $X = 1$, $Y = 1$. But suppose the values are $X_1 = 1 \cdot 06$, $Y_1 = 1 \cdot 02$ for the first cell and $X_2 = 0 \cdot 94$, $Y_2 = 0 \cdot 98$ for the second. Then the two morphogens will be being produced by chemical action at the rates $0 \cdot 18$, $0 \cdot 22$ respectively in the first cell and destroyed at the same rates in the second. At the same time there is a flow due to diffusion from the first cell to the second at the rate $0 \cdot 06$ for the first morphogen and $0 \cdot 18$ for the second. In sum the effect is a flow from the second cell to the first at the

[6]

rates 0·12, 0·04 for the two morphogens respectively. This flow tends to accentuate the already existing differences between the two cells. More generally, if

$$X_1 = 1+3\xi, \quad X_2 = 1-3\xi, \quad Y_1 = 1+\xi, \quad Y_2 = 1-\xi,$$

at some moment the four concentrations continue afterwards to be expressible in this form, and ξ increases at the rate 2ξ. Thus there is an exponential drift away from the equilibrium condition. It will be appreciated that a drift away from the equilibrium occurs with almost any small displacement from the equilibrium condition, though not normally according to an exact exponential curve. A particular choice was made in the above argument in order to exhibit the drift with only very simple mathematics.

Before it can be said to follow that a two-cell system can be unstable, with inhomogeneity succeeding homogeneity, it is necessary to show that the reaction rate functions postulated really can occur. To specify actual substances, concentrations and temperatures giving rise to these functions would settle the matter finally, but would be difficult and somewhat out of the spirit of the present inquiry. Instead, it is proposed merely to mention imaginary reactions which give rise to the required functions by the law of mass action, if suitable reaction constants are assumed. It will be sufficient to describe

(i) A set of reactions producing the first morphogen at the constant rate 1, and a similar set forming the second morphogen at the same rate.

(ii) A set destroying the second morphogen (Y) at the rate $7Y$.

(iii) A set converting the first morphogen (X) into the second (Y) at the rate $6X$.

(iv) A set producing the first morphogen (X) at the rate $11X$.

(v) A set destroying the first morphogen (X) at the rate $6Y$, so long as any of it is present.

The conditions of (i) can be fulfilled by reactions of the type $A \to X$, $B \to Y$, where A and B are substances continually present in large and invariable concentrations. The conditions of (ii) are satisfied by a reaction of the form $Y \to D$, D being an inert substance and (iii) by the reaction $X \to Y$ or $X \to Y + E$. The remaining two sets are rather more difficult. To satisfy the conditions of (iv) one may suppose that X is a catalyst for its own formation from A. The actual reactions could be the formation of an unstable compound U by the reaction $A + X \to U$, and the subsequent almost instantaneous breakdown $U \to 2X$. To destroy X at a rate proportional to Y as required in (v) one may suppose that a catalyst C is present in small but constant concentration and immediately combines with X, $X + C \to V$. The modified catalyst reacting with Y, at a rate proportional to Y, restores the catalyst but not the morphogen X, by the reactions $V + Y \to W$, $W \to C + H$, of which the latter is assumed instantaneous.

It should be emphasized that the reactions here described are by no means those which are most likely to give rise to instability in nature. The choice of the reactions to be discussed was dictated entirely by the fact that it was desirable that the argument be easy to follow. More plausible reaction systems are described in § 10.

Unstable equilibrium is not, of course, a condition which occurs very naturally. It usually requires some rather artificial interference, such as placing a marble on the top of a dome. Since systems tend to leave unstable equilibria they cannot often be in them. Such equilibria can, however, occur naturally through a stable equilibrium changing into an unstable one. For example, if a rod is hanging from a point a little above its centre of gravity it will be in stable equilibrium. If, however, a mouse climbs up the rod the equilibrium eventually becomes unstable and the rod starts to swing. A chemical analogue of this mouse-and-pendulum system would be that described above with the same diffusibilities but with the two morphogens produced at the rates

$$(3+I)X - 6Y + I - 1 \quad \text{and} \quad 6X - (9+I)Y - I + 1.$$

This system is stable if $I < 0$ but unstable if $I > 0$. If I is allowed to increase, corresponding to the mouse running up the pendulum, it will eventually become positive and the equilibrium will collapse. The system which was originally discussed was the case $I = 2$, and might be supposed to correspond to the mouse somehow reaching the top of the pendulum without disaster, perhaps by falling vertically on to it.

5. Left-handed and right-handed organisms

The object of this section is to discuss a certain difficulty which might be thought to show that the morphogen theory of morphogenesis cannot be right. The difficulty is mainly concerned with organisms which have not got bilateral symmetry. The argument, although carried through here without the use of mathematical formulae, may be found difficult by non-mathematicians, and these are therefore recommended to ignore it unless they are already troubled by such a difficulty.

An organism is said to have 'bilateral symmetry' if it is identical with its own reflexion in some plane. This plane of course always has to pass through some part of the organism, in particular through its centre of gravity. For the purpose of this argument it is more general to consider what may be called 'left-right symmetry'. An organism has left-right symmetry if its description in any right-handed set of rectangular Cartesian co-ordinates is identical with its description in some set of left-handed axes. An example of a body with left-right symmetry, but not bilateral symmetry, is a cylinder with the letter P printed on one end, and with the mirror image of a P on the other end, but with the two upright strokes of the two letters not parallel. The distinction may possibly be without a difference so far as the biological world is concerned, but mathematically it should not be ignored.

If the organisms of a species are sufficiently alike, and the absence of left-right symmetry sufficiently pronounced, it is possible to describe each individual as either right-handed or left-handed without there being difficulty in classifying any particular specimen. In man, for instance, one could take the X-axis in the forward direction, the Y-axis at right angles to it in the direction towards the side on which the heart is felt, and the Z-axis upwards. The specimen is classed as left-handed or right-handed according as the axes so chosen are left-handed or right-handed. A new classification has of course to be defined for each species.

The fact that there exist organisms which do not have left-right symmetry does not in itself cause any difficulty. It has already been explained how various kinds of symmetry can be lost in the development of the embryo, due to the particular disturbances (or 'noise') influencing the particular specimen not having that kind of symmetry, taken in conjunction with appropriate kinds of instability. The difficulty lies in the fact that there are species in which the proportions of left-handed and right-handed types are very unequal. It will be as well to describe first an argument which appears to show that this should not happen.

The argument is very general, and might be applied to a very wide class of theories of morphogenesis.

An entity may be described as '*P*-symmetrical' if its description in terms of one set of right-handed axes is identical with its description in terms of any other set of right-handed axes with the same origin. Thus, for instance, the totality of positions that a corkscrew would take up when rotated in all possible ways about the origin has *P*-symmetry. The entity will be said to be '*F*-symmetrical' when changes from right-handed axes to left-handed may also be made. This would apply if the corkscrew were replaced by a bilaterally symmetrical object such as a coal scuttle, or a left-right symmetrical object. In these terms one may say that there are species such that the totality of specimens from that species, together with the rotated specimens, is *P*-symmetrical, but very far from *F*-symmetrical. On the other hand, it is reasonable to suppose that

(i) The laws of physics are *F*-symmetrical.

(ii) The initial totality of zygotes for the species is *F*-symmetrical.

(iii) The statistical distribution of disturbances is *F*-symmetrical. The individual disturbances of course will in general have neither *F*-symmetry nor *P*-symmetry.

It should be noticed that the ideas of *P*-symmetry and *F*-symmetry as defined above apply even to so elaborate an entity as 'the laws of physics'. It should also be understood that the laws are to be the laws taken into account in the theory in question rather than some ideal as yet undiscovered laws.

Now it follows from these assumptions that the statistical distribution of resulting organisms will have *F*-symmetry, or more strictly that the distribution deduced as the result of working out such a theory will have such symmetry. The distribution of observed mature organisms, however, has no such symmetry In the first place, for instance, men are more often found standing on their feet than their heads. This may be corrected by taking gravity into account in the laws, together with an appropriate change of definition of the two kinds of symmetry. But it will be more convenient if, for the sake of argument, it is imagined that some species has been reared in the absence of gravity, and that the resulting distribution of mature organisms is found to be *P*-symmetrical but to yield more right-handed specimens than left-handed and so not to have *F*-symmetry. It remains therefore to explain this absence of *F*-symmetry.

Evidently one or other of the assumptions (i) to (iii) must be wrong, i.e. in a correct theory one of them would not apply. In the morphogen theory already described these three assumptions do all apply, and it must therefore be regarded as defective to some extent. The theory may be corrected by taking into account the fact that the morphogens do not always have an equal number of left- and right-handed molecules. According to one's point of view one may regard this as invalidating either (i), (ii) or even (iii). Simplest perhaps is to say that the totality of zygotes just is not *F*-symmetrical, and that this could be seen if one looked at the molecules. This is, however, not very satisfactory from the point of view of this paper, as it would not be consistent with describing states in terms of concentrations only. It would be preferable if it was found possible to find more accurate laws concerning reactions and diffusion. For the purpose of accounting for unequal numbers of left- and right-handed organisms it is unnecessary to do more than show that there are corrections which would not be *F*-symmetrical when there are laevo- or dextrorotatory morphogens, and which would be large enough to account for the effects observed. It is not very difficult to think of such effects. They do not have to be very large, but must, of course, be larger than the purely statistical effects, such as thermal noise or Brownian movement.

There may also be other reasons why the totality of zygotes is not F-symmetrical, e.g. an asymmetry of the chromosomes themselves. If these also produce a sufficiently large effect, so much the better.

Though these effects may be large compared with the statistical disturbances they are almost certainly small compared with the ordinary diffusion and reaction effects. This will mean that they only have an appreciable effect during a short period in which the breakdown of left-right symmetry is occurring. Once their existence is admitted, whether on a theoretical or experimental basis, it is probably most convenient to give them mathematical expression by regarding them as P-symmetrically (but not F-symmetrically) distributed disturbances. However, they will not be considered further in this paper.

6. REACTIONS AND DIFFUSION IN A RING OF CELLS

The original reason for considering the breakdown of homogeneity was an apparent difficulty in the diffusion-reaction theory of morphogenesis. Now that the difficulty is resolved it might be supposed that there is no reason for pursuing this aspect of the problem further, and that it would be best to proceed to consider what occurs when the system is very far from homogeneous. A great deal more attention will nevertheless be given to the breakdown of homogeneity. This is largely because the assumption that the system is still nearly homogeneous brings the problem within the range of what is capable of being treated mathematically. Even so many further simplifying assumptions have to be made. Another reason for giving this phase such attention is that it is in a sense the most critical period. That is to say, that if there is any doubt as to how the organism is going to develop it is conceivable that a minute examination of it just after instability has set in might settle the matter, but an examination of it at any earlier time could never do so.

There is a great variety of geometrical arrangement of cells which might be considered, but one particular type of configuration stands out as being particularly simple in its theory, and also illustrates the general principles very well. This configuration is a ring of similar cells. One may suppose that there are N such cells. It must be admitted that there is no biological example to which the theory of the ring can be immediately applied, though it is not difficult to find ones in which the principles illustrated by the ring apply.

It will be assumed at first that there are only two morphogens, or rather only two interesting morphogens. There may be others whose concentration does not vary either in space or time, or which can be eliminated from the discussion for one reason or another. These other morphogens may, for instanse, be catalysts involved in the reactions between the interesting morphogens. An example of a complete system of reactions is given in § 10. Some consideration will also be given in §§ 8, 9 to the case of three morphogens. The reader should have no difficulty in extending the results to any number of morphogens, but no essentially new features appear when the number is increased beyond three.

The two morphogens will be called X and Y. These letters will also be used to denote their concentrations. This need not lead to any real confusion. The concentration of X in cell r may be written X_r, and Y_r has a similar meaning. It is convenient to regard 'cell N' and 'cell O' as synonymous, and likewise 'cell 1' and cell '$N+1$'. One can then say that for each r satisfying $1 \leqslant r \leqslant N$ cell r exchanges material by diffusion with cells $r-1$ and $r+1$. The cell-to-cell diffusion constant for X will be called μ, and that for Y will be called ν. This means that for unit concentration difference of X, this morphogen passes at the rate μ from the cell with the higher concentration to the (neighbouring) cell with the lower concentration. It is also necessary to make assumptions about the rates of chemical reaction. The most general assumption that can be made is that for concentrations X and Y chemical reactions are tending to increase X at the rate $f(X, Y)$ and Y at the rate $g(X, Y)$. When the changes in X and Y due to diffusion are also taken into account the behaviour of the system may be described by the $2N$ differential equations

$$\left.\begin{aligned} \frac{\mathrm{d}X_r}{\mathrm{d}t} &= f(X_r, Y_r) + \mu(X_{r+1} - 2X_r + X_{r-1}) \\ \frac{\mathrm{d}Y_r}{\mathrm{d}t} &= g(X_r, Y_r) + \nu(Y_{r+1} - 2Y_r + Y_{r-1}) \end{aligned}\right\} \quad (r = 1, ..., N). \qquad (6\cdot1)$$

If $f(h, k) : g(h, k) = 0$, then an isolated cell has an equilibrium with concentrations $X = h$, $Y = k$. The ring system also has an equilibrium, stable or unstable, with each X_r equal to h and each Y_r equal to k. Assuming that the system is not very far from this equilibrium it is convenient to put $X_r = h + x_r$, $Y_r = k + y_r$. One may also write $ax + by$ for $f(h+x, y+k)$ and $cx + dy$ for $g(h+x, y+k)$. Since $f(h, k) = g(h, k) = 0$ no constant terms are required, and since x and y are supposed small the terms in higher powers of x and y will have relatively little effect and one is justified in ignoring them. The four quantities a, b, c, d may be called the 'marginal reaction rates'. Collectively they may be described as the 'marginal reaction rate matrix'. When there are M morphogens this matrix consists of M^2 numbers. A marginal reaction rate has the dimensions of the reciprocal of a time, like a radioactive decay rate, which is in fact an example of a marginal (nuclear) reaction rate.

With these assumptions the equations can be rewritten as

$$\left.\begin{aligned} \frac{\mathrm{d}x_r}{\mathrm{d}t} &= ax_r + by_r + \mu(x_{r+1} - 2x_r + x_{r-1}), \\ \frac{\mathrm{d}y_r}{\mathrm{d}t} &= cx_r + dy_r + \nu(y_{r+1} - 2y_r + y_{r-1}). \end{aligned}\right\} \qquad (6\cdot2)$$

To solve the equations one introduces new co-ordinates $\xi_0, ..., \xi_{N-1}$ and $\eta_0, ..., \eta_{N-1}$ by putting

$$\left.\begin{aligned} x_r &= \sum_{s=0}^{N-1} \exp\left[\frac{2\pi irs}{N}\right] \xi_s, \\ y_r &= \sum_{s=0}^{N-1} \exp\left[\frac{2\pi irs}{N}\right] \eta_s. \end{aligned}\right\} \qquad (6\cdot3)$$

These relations can also be written as

$$\left.\begin{aligned} \xi_r &= \frac{1}{N} \sum_{s=1}^{N} \exp\left[-\frac{2\pi irs}{N}\right] x_s, \\ \eta_r &= \frac{1}{N} \sum_{s=1}^{N} \exp\left[-\frac{2\pi irs}{N}\right] y_s. \end{aligned}\right\} \qquad (6\cdot4)$$

6-2

as may be shown by using the equations

$$\sum_{s=1}^{N} \exp\left[\frac{2\pi i r s}{N}\right] = 0 \quad \text{if} \quad 0 < r < N,$$
$$= N \quad \text{if} \quad r = 0 \text{ or } r = N, \tag{6.5}$$

(referred to in § 2). Making this substitution one obtains

$$\frac{d\xi_s}{dt} = \frac{1}{N} \sum_{s=1}^{N} \exp\left[-\frac{2\pi i r s}{N}\right] \left[ax_r + by_r + \mu\left(\exp\left[-\frac{2\pi i s}{N}\right] - 2 + \exp\left[\frac{2\pi i s}{N}\right]\right)\xi_s\right]$$

$$= a\xi_s + b\eta_s + \mu\left(\exp\left[-\frac{2\pi i s}{N}\right] - 2 + \exp\left[\frac{2\pi i s}{N}\right]\right)\xi_s$$

$$= \left(a - 4\mu\sin^2\frac{\pi s}{N}\right)\xi_s + b\eta_s. \tag{6.6}$$

Likewise
$$\frac{d\eta_s}{dt} = c\xi_s + \left(d - 4\nu\sin^2\frac{\pi s}{N}\right)\eta_s. \tag{6.7}$$

The equations have now been converted into a quite manageable form, with the variables separated. There are now two equations concerned with ξ_1 and η_1, two concerned with ξ_2 and η_2, etc. The equations themselves are also of a well-known standard form, being linear with constant coefficients. Let p_s and p'_s be the roots of the equation

$$\left(p - a + 4\mu\sin^2\frac{\pi s}{N}\right)\left(p - d + 4\nu\sin^2\frac{\pi s}{N}\right) = bc \tag{6.8}$$

(with $\mathscr{R}p_s \geqslant \mathscr{R}p'_s$ for definiteness), then the solution of the equations is of the form

$$\left.\begin{aligned}\xi_s &= A_s e^{p_s t} + B_s e^{p'_s t},\\ \eta_s &= C_s e^{p_s t} + D_s e^{p'_s t},\end{aligned}\right\} \tag{6.9}$$

where, however, the coefficients A_s, B_s, C_s, D_s are not independent but are restricted to satisfy

$$\left.\begin{aligned}A_s\left(p_s - a + 4\mu\sin^2\frac{\pi s}{N}\right) &= bC_s,\\ B_s\left(p'_s - a + 4\mu\sin^2\frac{\pi s}{N}\right) &= bD_s.\end{aligned}\right\} \tag{6.10}$$

If it should happen that $p_s = p'_s$ the equations (6.9) have to be replaced by

$$\left.\begin{aligned}\xi_s &= (A_s + B_s t) e^{p_s t},\\ \eta_s &= (C_s + D_s t) e^{p_s t}.\end{aligned}\right\} \tag{6.9}'$$

and (6.10) remains true. Substituting back into (6.3) and replacing the variables x_r, y_r by X_r, Y_r (the actual concentrations) the solution can be written

$$\left.\begin{aligned}X_r &= h + \sum_{s=1}^{N} (A_s e^{p_s t} + B_s e^{p'_s t}) \exp\left[\frac{2\pi i r s}{N}\right],\\ Y_r &= k + \sum_{s=1}^{N} (C_s e^{p_s t} + D_s e^{p'_s t}) \exp\left[\frac{2\pi i r s}{N}\right].\end{aligned}\right\} \tag{6.11}$$

Here A_s, B_s, C_s, D_s are still related by (6.10), but otherwise are arbitrary complex numbers; p_s and p'_s are the roots of (6.8).

The expression (6·11) gives the general solution of the equations (6·1) when one assumes that departures from homogeneity are sufficiently small that the functions $f(X, Y)$ and $g(X, Y)$ can safely be taken as linear. The form (6·11) given is not very informative. It will be considerably simplified in § 8. Another implicit assumption concerns random disturbing influences. Strictly speaking one should consider such influences to be continuously at work. This would make the mathematical treatment considerably more difficult without substantially altering the conclusions. The assumption which is implicit in the analysis, here and in § 8, is that the state of the system at $t = 0$ is not one of homogeneity, since it has been displaced from such a state by the disturbances; but after $t = 0$ further disturbances are ignored. In § 9 the theory is reconsidered without this latter assumption.

7. CONTINUOUS RING OF TISSUE

As an alternative to a ring of separate cells one might prefer to consider a continuous ring of tissue. In this case one can describe the position of a point of the ring by the angle θ which a radius to the point makes with a fixed reference radius. Let the diffusibilities of the two substances be μ' and ν'. These are not quite the same as μ and ν of the last section, since μ and ν are in effect referred to a cell diameter as unit of length, whereas μ' and ν' are referred to a conventional unit, the same unit in which the radius ρ of the ring is measured. Then

$$\mu = \mu'\left(\frac{N}{2\pi\rho}\right)^2, \quad \nu = \nu'\left(\frac{N}{2\pi\rho}\right)^2.$$

The equations are
$$\begin{aligned}
\frac{\partial X}{\partial t} &= a(X-h)+b(Y-k)+\frac{\mu'}{\rho^2}\frac{\partial^2 X}{\partial\theta^2}, \\
\frac{\partial Y}{\partial t} &= c(X-h)+d(Y-k)+\frac{\nu'}{\rho^2}\frac{\partial^2 Y}{\partial\theta^2},
\end{aligned} \quad (7\cdot1)$$

which will be seen to be the limiting case of (6·2). The marginal reaction rates a, b, c, d are, as before, the values at the equilibrium position of $\partial f/\partial X$, $\partial f/\partial Y$, $\partial g/\partial X$, $\partial g/\partial Y$. The general solution of the equations is

$$\begin{aligned}
X &= h+ \sum_{s=-\infty}^{\infty} (A_s\, e^{p_s t}+B_s\, e^{p_s' t})\, e^{is\theta}, \\
Y &= k+ \sum_{s=-\infty}^{\infty} (C_s\, e^{p_s t}+D_s\, e^{p_s' t})\, e^{is\theta},
\end{aligned} \quad (7\cdot2)$$

where p_s, p_s' are now roots of

$$\left(p-a+\frac{\mu's^2}{\rho^2}\right)\left(p-d+\frac{\nu's^2}{\rho^2}\right) = bc \quad (7\cdot3)$$

and
$$\begin{aligned}
A_s\left(p_s-a+\frac{\mu's^2}{\rho^2}\right) &= bC_s, \\
B_s\left(p_s'-a+\frac{\mu's^2}{\rho^2}\right) &= bD_s.
\end{aligned} \quad (7\cdot4)$$

This solution may be justified by considering the limiting case of the solution (6·11). Alternatively, one may observe that the formula proposed is a solution, so that it only remains to prove that it is the most general one. This will follow if values of A_s, B_s, C_s, D_s can be found to fit any given initial conditions. It is well known that any function of an angle (such as X) can be expanded as a 'Fourier series'

$$X(\theta) = \sum_{s=-\infty}^{\infty} G_s e^{is\theta} \quad (X(\theta) \text{ being values of } X \text{ at } t = 0),$$

provided, for instance, that its first derivative is continuous. If also

$$Y(\theta) = \sum_{s=-\infty}^{\infty} H_s e^{is\theta} \quad (Y(\theta) \text{ being values of } Y \text{ at } t = 0),$$

then the required initial conditions are satisfied provided $A_s + B_s = G_s$ and $C_s + D_s = H_s$. Values A_s, B_s, C_s, D_s to satisfy these conditions can be found unless $p_s = p'_s$. This is an exceptional case and its solution if required may be found as the limit of the normal case.

8. Types of asymptotic behaviour in the ring after a lapse of time

As the reader was reminded in §2, after a lapse of time the behaviour of an expression of the form of (6·11) is eventually dominated by the terms for which the corresponding p_s has the largest real part. There may, however, be several terms for which this real part has the same value, and these terms will together dominate the situation, the other terms being ignored by comparison. There will, in fact, normally be either two or four such 'leading' terms. For if p_{s_0} is one of them then $p_{N-s_0} = p_{s_0}$, since

$$\sin^2 \frac{\pi(N-s_0)}{N} = \sin^2 \frac{\pi s_0}{N},$$

so that p_{s_0} and p_{N-s_0} are roots of the same equation (6·8). If also p_{s_0} is complex then $\mathscr{R}p_{s_0} = \mathscr{R}p'_{s_0}$, and so in all

$$\mathscr{R}p_{s_0} = \mathscr{R}p'_{s_0} = \mathscr{R}p_{N-s_0} = \mathscr{R}p'_{N-s_0}.$$

One need not, however, normally anticipate that any further terms will have to be included. If p_{s_0} and p_{s_1} are to have the same real part, then, unless $s_1 = s_0$ or $s_0 + s_1 = N$ the quantities a, b, c, d, μ, ν will be restricted to satisfy some special condition, which they would be unlikely to satisfy by chance. It is possible to find circumstances in which as many as ten terms have to be included if such special conditions *are* satisfied, but these have no particular physical or biological importance. It is assumed below that none of these chance relations hold.

It has already been seen that it is necessary to distinguish the cases where the value of p_{s_0} for one of the dominant terms is real from those where it is complex. These may be called respectively the *stationary* and the *oscillatory* cases.

Stationary case. After a sufficient lapse of time $X_r - h$ and $Y_r - k$ approach asymptotically to the forms

$$\left.\begin{aligned} X_r - h &= 2\mathscr{R}A_{s_0} \exp\left[\frac{2\pi i s_0 r}{N} + It\right], \\ Y_r - k &= 2\mathscr{R}C_{s_0} \exp\left[\frac{2\pi i s_0 r}{N} + It\right]. \end{aligned}\right\} \tag{8·1}$$

Oscillatory case. After a sufficient lapse of time $X_r - h$ and $Y_r - k$ approach the forms

$$\left.\begin{aligned} X_r - h &= 2e^{It}\mathscr{R}\left\{A_{s_0}\exp\left[\frac{2\pi i s_0 r}{N} + i\omega t\right] + A_{N-s_0}\exp\left[-\frac{2\pi i s_0 r}{N} - i\omega t\right]\right\}, \\ Y_r - k &= 2e^{It}\mathscr{R}\left\{C_{s_0}\exp\left[\frac{2\pi i s_0 r}{N} + i\omega t\right] + C_{N-s_0}\exp\left[-\frac{2\pi i s_0 r}{N} - i\omega t\right]\right\}. \end{aligned}\right\} \tag{8·2}$$

[[14]]

The real part of p_{s_0} has been represented by I, standing for 'instability', and in the oscillatory case its imaginary part is ω. By the use of the $\mathscr{R}$ operation (real part of), two terms have in each case been combined in one.

The meaning of these formulae may be conveniently described in terms of waves. In the stationary case there are stationary waves on the ring having s_0 lobes or crests. The coefficients A_{s_0} and C_{s_0} are in a definite ratio given by $(6\cdot10)$, so that the pattern for one morphogen determines that for the other. With the lapse of time the waves become more pronounced provided there is genuine instability, i.e. if I is positive. The wave-length of the waves may be obtained by dividing the number of lobes into the circumference of the ring. In the oscillatory case the interpretation is similar, but the waves are now not stationary but travelling. As well as having a wave-length they have a velocity and a frequency. The frequency is $\omega/2\pi$, and the velocity is obtained by multiplying the wave-length by the frequency. There are two wave trains moving round the ring in opposite directions.

The wave-lengths of the patterns on the ring do not depend only on the chemical data a, b, c, d, μ', ν' but on the circumference of the ring, since they must be submultiples of the latter. There is a sense, however, in which there is a 'chemical wave-length' which does not depend on the dimensions of the ring. This may be described as the limit to which the wave-lengths tend when the rings are made successively larger. Alternatively (at any rate in the case of continuous tissue), it may be described as the wave-length when the radius is chosen to give the largest possible instability I. One may picture the situation by supposing that the chemical wave-length is true wave-length which is achieved whenever possible, but that on a ring it is necessary to 'make do' with an approximation which divides exactly into the circumference.

Although all the possibilities are covered by the stationary and oscillatory alternatives there are special cases of them which deserve to be treated separately. One of these occurs when $s_0 = 0$, and may be described as the 'case of extreme long wave-length', though this term may perhaps preferably be reserved to describe the chemical data when they are such that s_0 is zero whatever the dimensions of the ring. There is also the case of 'extreme short wave-length'. This means that $\sin^2 (\pi s_0/N)$ is as large as possible, which is achieved by s_0 being either $\frac{1}{2}N$, or $\frac{1}{2}(N-1)$. If the remaining possibilities are regarded as forming the 'case of finite wave-length', there are six subcases altogether. It will be shown that each of these really can occur, although two of them require three or more morphogens for their realization.

(a) *Stationary case with extreme long wave-length.* This occurs for instance if $\mu = \nu = \frac{1}{4}$, $b = c = 1$, $a = d$. Then $p_s = a - \sin^2 \dfrac{\pi s}{N} + 1$. This is certainly real and is greatest when $s = 0$.

In this case the contents of all the cells are the same; there is no resultant flow from cell to cell due to diffusion, so that each is behaving as if it were isolated. Each is in unstable equilibrium, and slips out of it in synchronism with the others.

(b) *Oscillatory case with extreme long wave-length.* This occurs, for instance, if $\mu = \nu = \frac{1}{4}$, $b = -c = 1$, $a = d$. Then $p_s = a - \sin^2 \dfrac{\pi s}{N} \pm i$. This is complex and its real part is greatest when $s = 0$. As in case (a) each cell behaves as if it were isolated. The difference from case (a) is that the departure from the equilibrium is oscillatory.

(c) *Stationary waves of extreme short wave-length.* This occurs, for instance, if $\nu = 0$, $\mu = 1$, $d = I$, $a = I - 1$, $b = -c = 1$. p_s is

$$I - \tfrac{1}{2} - 2\sin^2\frac{\pi s}{N} + \sqrt{\left\{\left(2\sin^2\frac{\pi s}{N} + \frac{1}{2}\right)^2 - 1\right\}},$$

and is greatest when $\sin^2(\pi s/N)$ is greatest. If N is even the contents of each cell are similar to those of the next but one, but distinctly different from those of its immediate neighbours. If, however, the number of cells is odd this arrangement is impossible, and the magnitude of the difference between neighbouring cells varies round the ring, from zero at one point to a maximum at a point diametrically opposite.

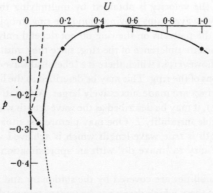

FIGURE 1. Values of $\mathscr{R}p$ (instability or growth rate), and $|\mathscr{I}p|$ (radian frequency of oscillation), related to wave-length $2\pi U^{-\frac{1}{2}}$ as in the relation (8·3) with $I = 0$. This is a case of stationary waves with finite wave-length. Full line, $\mathscr{R}p$; broken line, $-|\mathscr{I}p|$ (zero for $U > 0\cdot071$); dotted line, $\mathscr{R}p'$. The full circles on the curve for $\mathscr{R}p$ indicate the values of U, p actually achievable on the finite ring considered in §10, with $s = 0$ on the extreme left, $s = 5$ on the right.

(d) *Stationary waves of finite wave-length.* This is the case which is of greatest interest, and has most biological application. It occurs, for instance, if $a = I - 2$, $b = 2\cdot5$, $c = -1\cdot25$, $d = I + 1\cdot5$, $\mu' = 1$, $\nu' = \tfrac{1}{2}$, and $\dfrac{\mu}{\mu'} = \dfrac{\nu}{\nu'} = \left(\dfrac{N}{2\pi\rho}\right)^2$. As before ρ is the radius of the ring, and N the number of cells in it. If one writes U for $\left(\dfrac{N}{\pi\rho}\right)^2\sin^2\dfrac{\pi s}{N}$, then equation (6·8) can, with these special values, be written

$$(p - I)^2 + (\tfrac{1}{2} + \tfrac{3}{2}U)(p - I) + \tfrac{1}{2}(U - \tfrac{1}{2})^2 = 0. \tag{8.3}$$

This has a solution $p = I$ if $U = \tfrac{1}{2}$. On the other hand, it will be shown that if U has any other (positive) value then both roots for $p - I$ have negative real parts. Their product is positive being $\tfrac{1}{2}(U - \tfrac{1}{2})^2$, so that if they are real they both have the same sign. Their sum in this case is $-\tfrac{1}{2} - \tfrac{3}{2}U$ which is negative. Their common sign is therefore negative. If, however, the roots are complex their real parts are both equal to $-\tfrac{1}{4} - \tfrac{3}{4}U$, which is negative.

If the radius ρ of the ring be chosen so that for some integer s_0, $\frac{1}{2} = U = \left(\frac{N}{\pi\rho}\right)^2 \sin^2 \frac{\pi s_0}{N}$, there will be stationary waves with s_0 lobes and a wave-length which is also equal to the chemical wave-length, for p_{s_0} will be equal to I, whereas every other p_s will have a real part smaller than I. If, however, the radius is chosen so that $\left(\frac{N}{\pi\rho}\right)^2 \sin^2 \frac{\pi s}{N} = \frac{1}{2}$ cannot hold with an integral s, then (in this example) the actual number of lobes will be one of the two integers nearest to the (non-integral) solutions of this equation, and usually *the* nearest. Examples can, however, be constructed where this simple rule does not apply.

Figure 1 shows the relation (8·3) in graphical form. The curved portions of the graphs are hyperbolae.

The two remaining possibilities can only occur with three or more morphogens. With one morphogen the only possibility is (a).

(e) *Oscillatory case with a finite wave-length.* This means that there are genuine travelling waves. Since the example to be given involves three morphogens it is not possible to use the formulae of § 6. Instead, one must use the corresponding three morphogen formulae. That which corresponds to (6·8) or (7·3) is most conveniently written as

$$\begin{vmatrix} a_{11}-p-\mu_1 U & a_{12} & a_{13} \\ a_{21} & a_{22}-p-\mu_2 U & a_{23} \\ a_{31} & a_{32} & a_{33}-p-\mu_3 U \end{vmatrix} = 0, \qquad (8\cdot4)$$

where again U has been written for $\left(\frac{N}{\pi\rho}\right)^2 \sin^2 \frac{\pi s}{N}$. (This means essentially that $U = \left(\frac{2\pi}{\lambda}\right)^2$, where λ is the wave-length.) The four marginal reactivities are superseded by nine $a_{11}, ..., a_{33}$, and the three diffusibilities are μ_1, μ_2, μ_3. Special values leading to travelling waves are

$$\left.\begin{array}{lll} \mu_1 = \frac{2}{3}, & \mu_2 = \frac{1}{3}, & \mu_3 = 0 \\ a_{11} = -\frac{10}{3}, & a_{12} = 3, & a_{13} = -1, \\ a_{21} = -2, & a_{22} = \frac{7}{3}, & a_{23} = 0, \\ a_{31} = 3, & a_{32} = -4, & a_{33} = 0, \end{array}\right\} \qquad (8\cdot5)$$

and with them (8·4) reduces to

$$p^3 + p^2(U+1) + p(1 + \tfrac{2}{3}(U-1)^2) + U + 1 = 0. \qquad (8\cdot6)$$

If $U = 1$ the roots are $\pm i$ and -2. If U is near to I they are approximately $-1-U$ and $\pm i + \frac{(U-1)^2}{18}(\pm i - 1)$, and all have negative real parts. If the greatest real part is not the value zero, achieved with $U = 1$, then the value zero must be reached again at some intermediate value of U. Since P is then pure imaginary the even terms of (8·6) must vanish, i.e. $(p^2+1)(U+1) = 0$. But this can only happen if $p = \pm i$, and the vanishing of the odd terms then shows that $U = 1$. Hence zero is the largest real part for any root p of (8·6). The corresponding p is $\pm i$ and U is 1. This means that there are travelling waves with unit (chemical) radian frequency and unit (chemical) velocity. If I is added to a_{11}, a_{22} and a_{33}, the instability will become I in place of zero.

[17]

(*f*) Oscillatory case with extreme short wave-length. This means that there is metabolic oscillation with neighbouring cells nearly 180° out of phase. It can be achieved with three morphogens and the following chemical data:

$$\left.\begin{aligned} \mu = 1, \qquad \mu_2 &= \mu_3 = 0, \\ a_{11} = -1, \quad a_{12} &= -1, \quad a_{13} = 0, \\ a_{21} = 1, \quad a_{22} &= 0, \quad a_{23} = -1, \\ a_{31} = 0, \quad a_{32} &= 1, \quad a_{33} = 0. \end{aligned}\right\} \tag{8.7}$$

With these values (8·4) reduces to

$$p^3 + p^2(U+1) + 2p + U + 1 = 0. \tag{8.8}$$

This may be shown to have all the real parts of its roots negative if $U \geqslant 0$, for if $U = 0$ the roots are near to -0.6, $-0.2 \pm 1.3i$, and if U be continuously increased the values of p will alter continuously. If they ever attain values with a positive real part they must pass through pure imaginary values (or zero). But if p is pure imaginary $p^3 + 2p$ and $(p^2 + 1)(U + 1)$ must both vanish, which is impossible if $U \geqslant 0$. As U approaches infinity, however, one of the roots approaches i. Thus $\mathscr{R}p = 0$ can be approached as closely as desired by large values of U, but not attained.

9. Further consideration of the mathematics of the ring

In this section some of the finer points concerning the development of wave patterns are considered. These will be of interest mainly to those who wish to do further research on the subject, and can well be omitted on a first reading.

(1) *General formulae for the two morphogen case.* Taking the limiting case of a ring of large radius (or a filament), one may write $\left(\dfrac{N}{\pi\rho}\right)^2 \sin^2 \dfrac{\pi s}{N} = U = \left(\dfrac{2\pi}{\lambda}\right)^2$ in (6·11) or $\dfrac{s^2}{\rho^2} = U = \left(\dfrac{2\pi}{\lambda}\right)^2$ in (7·3) and obtain

$$(p - a + \mu' U)(p - d + \nu' U) = bc, \tag{9.1}$$

which has the solution

$$p = \frac{a+d}{2} - \frac{\mu' + \nu'}{2} U \pm \sqrt{\left\{\left(\frac{\mu' - \nu'}{2} U + \frac{d-a}{2}\right)^2 + bc\right\}}. \tag{9.2}$$

One may put $I(U)$ for the real part of this, representing the instability for waves of wave-length $\lambda = 2\pi U^{-\frac{1}{2}}$. The dominant waves correspond to the maximum of $I(U)$. This maximum may either be at $U = 0$ or $U = \infty$ or at a stationary point on the part of the curve which is hyperbolic (rather than straight). When this last case occurs the values of p (or I) and U at the maximum are

$$\left.\begin{aligned} p = I &= (d\mu' - a\nu' - 2\sqrt{(\mu'\nu')}\sqrt{(-bc)})(\mu' - \nu')^{-1}, \\ U &= \left(a - d + \frac{\mu' + \nu'}{\sqrt{(\mu'\nu')}}\sqrt{(-bc)}\right)(\mu' - \nu')^{-1}. \end{aligned}\right\} \tag{9.3}$$

The conditions which lead to the four cases (*a*), (*b*), (*c*), (*d*) described in the last section are

(*a*) (Stationary waves of extreme long wave-length.) This occurs if either

(i) $bc > 0$, (ii) $bc < 0$ and $\dfrac{d-a}{\sqrt{(-bc)}} > \dfrac{\mu' + \nu'}{\sqrt{(\mu'\nu')}}$, (iii) $bc < 0$ and $\dfrac{d-a}{\sqrt{(-bc)}} < -2$.

The condition for instability in either case is that either $bc > ad$ or $a + d > 0$.

(b) (Oscillating case with extreme long wave-length, i.e. synchronized oscillations.) This occurs if

$$bc < 0 \quad \text{and} \quad -2 < \frac{d-a}{\sqrt{(-bc)}} < \frac{4\sqrt{(\mu'\nu')}}{\mu'+\nu'}.$$

There is instability if in addition $a+d > 0$.

(c) (Stationary waves of extreme short wave-length.) This occurs if $bc < 0$, $\mu' > \nu' = 0$. There is instability if, in addition, $a+d > 0$.

(d) (Stationary waves of finite wave-length.) This occurs if

$$bc < 0 \quad \text{and} \quad \frac{4\sqrt{(\mu'\nu')}}{\mu'+\nu'} < \frac{d-a}{\sqrt{(-bc)}} < \frac{\mu'+\nu'}{(\sqrt{\mu'\nu'})}, \tag{9.4a}$$

and there is instability if also

$$\frac{d}{\sqrt{(-bc)}} \sqrt{\frac{\mu'}{\nu'}} - \frac{a}{\sqrt{(-bc)}} \sqrt{\frac{\nu'}{\mu'}} > 2. \tag{9.4b}$$

It has been assumed that $\nu' \leqslant \mu' > 0$. The case where $\mu' \leqslant \nu' > 0$ can be obtained by interchanging the two morphogens. In the case $\mu' = \nu' = 0$ there is no co-operation between the cells whatever.

Some additional formulae will be given for the case of stationary waves of finite wavelength. The marginal reaction rates may be expressed parametrically in terms of the diffusibilities, the wave-length, the instability, and two other parameters α and χ. Of these α may be described as the ratio of $X-h$ to $Y-k$ in the waves. The expressions for the marginal reaction rates in terms of these parameters are

$$\left. \begin{array}{l} a = \mu'(\nu'-\mu')^{-1}(2\nu'U_0+\chi)+I, \\ b = \mu'(\nu'-\mu')^{-1}((\mu'+\nu')U_0+\chi)\alpha, \\ c = \nu'(\mu'-\nu')^{-1}((\mu'+\nu')U_0+\chi)\alpha^{-1}, \\ d = \nu'(\mu'-\nu')^{-1}(2\mu'U_0+\chi)+I, \end{array} \right\} \tag{9.5}$$

and when these are substituted into (9.2) it becomes

$$p = I - \tfrac{1}{2}\chi - \frac{\mu'+\nu'}{2}U + \sqrt{\left\{ \left(\frac{\mu'+\nu'}{2}U+\tfrac{1}{2}\chi\right)^2 - \mu'\nu'(U-U_0)^2 \right\}}. \tag{9.6}$$

Here $2\pi U_0^{-\frac{1}{2}}$ is the chemical wave-length and $2\pi U^{-\frac{1}{2}}$ the wave-length of the Fourier component under consideration. χ must be positive for case (d) to apply.

If s be regarded as a continuous variable one can consider (9.2) or (9.6) as relating s to p, and dp/ds and d^2p/ds^2 have meaning. The value of d^2p/ds^2 at the maximum is of some interest, and will be used below in this section. Its value is

$$\frac{d^2p}{ds^2} = -\frac{\sqrt{(\mu'\nu')}}{\rho^2} \cdot \frac{8\sqrt{(\mu'\nu')}}{\mu'+\nu'} \cos^2 \frac{\pi s}{N} (1+\chi U_0^{-1}(\mu'+\nu')^{-1})^{-1}. \tag{9.7}$$

(2) In §§ 6, 7, 8 it was supposed that the disturbances were not continuously operative, and that the marginal reaction rates did not change with the passage of time. These assumptions will now be dropped, though it will be necessary to make some other, less drastic,

7-2

approximations to replace them. The (statistical) amplitude of the 'noise' disturbances will be assumed constant in time. Instead of (6·6), (6·7), one then has

$$\frac{d\xi}{dt} = a'\xi + b\eta + R_1(t),$$
$$\frac{d\eta}{dt} = c\xi + d'\eta + R_2(t),$$

$$(9·8)$$

where ξ, η have been written for ξ_s, η_s since s may now be supposed fixed. For the same reason $a - 4\mu\sin^2\frac{\pi s}{N}$ has been replaced by a' and $d - 4\nu\sin^2\frac{\pi s}{N}$ by d'. The noise disturbances may be supposed to constitute white noise, i.e. if (t_1, t_2) and (t_3, t_4) are two non-overlapping intervals then $\int_{t_1}^{t_2} R_1(t)\,dt$ and $\int_{t_3}^{t_4} R_2(t)\,dt$ are statistically independent and each is normally distributed with variances $\beta_1(t_2 - t_1)$ and $\beta_1(t_4 - t_3)$ respectively, β_1 being a constant describing the amplitude of the noise. Likewise for $R_2(t)$, the constant β_1 being replaced by β_2. If p and p' are the roots of $(p - a')(p - d') = bc$ and p is the greater (both being real), one can make the substitution

$$\xi = b(u+v),$$
$$\eta = (p - a')u + (p' - a')v,$$

$$(9·9)$$

which transforms (9·8) into

$$\frac{du}{dt} = pu + \frac{p' - a'}{(p' - p)b}R_1(t) - \frac{R_2(t)}{p' - p} + \xi\frac{d}{dt}\left(\frac{p' - a'}{(p' - p)b}\right) - \eta\frac{d}{dt}\left(\frac{1}{p' - p}\right), \qquad (9·11)$$

with a similar equation for v, of which the leading terms are $dv/dt = p'v$. This indicates that v will be small, or at least small in comparison with u after a lapse of time. If it is assumed that $v = 0$ holds (9·11) may be written

$$\frac{du}{dt} = qu + L_1(t)R_1(t) + L_2(t)R_2(t), \qquad (9·12)$$

where

$$L_1(t) = \frac{p' - a'}{(p' - p)b}, \quad L_2(t) = \frac{1}{p' - p}, \quad q = p + bL_1'(t). \qquad (9·13)$$

The solution of this equation is

$$u = \int_{-\infty}^{t} (L_1(w)R_1(w) + L_2(w)R_2(w))\exp\left[\int_{w}^{t} q(z)\,dz\right]dw. \qquad (9·14)$$

One is, however, not so much interested in such a solution in terms of the statistical disturbances as in the consequent statistical distribution of values of u, ξ and η at various times after instability has set in. In view of the properties of 'white noise' assumed above, the values of u at time t will be distributed according to the normal error law, with the variance

$$\int_{-\infty}^{t} [\beta_1(L_1(w))^2 + \beta_2(L_2(w))^2]\exp\left[2\int_{w}^{t} q(z)\,dz\right]dw. \qquad (9·15)$$

There are two commonly occurring cases in which one can simplify this expression considerably without great loss of accuracy. If the system is in a distinctly stable state, then $q(t)$, which is near to $p(t)$, will be distinctly negative, and $\exp\left[\int_w^t q(z)\,dz\right]$ will be small unless w is near to t. But then $L_1(w)$ and $L_2(w)$ may be replaced by $L_1(t)$ and $L_2(t)$ in the integral, and also $q(z)$ may be replaced by $q(t)$. With these approximations the variance is

$$(-2q(t))^{-1}[\beta_1(L_1(t))^2+\beta_2(L_2(t))^2]. \tag{9.16}$$

A second case where there is a convenient approximation concerns times when the system is unstable, so that $q(t) > 0$. For the approximation concerned to apply $2\int_w^t q(z)\,dz$ must have its maximum at the last moment $w(= t_0)$ when $q(t_0) = 0$, and it must be the maximum by a considerable margin (e.g. at least 5) over all other local maxima. These conditions would apply for instance if $q(z)$ were always increasing and had negative values at a sufficiently early time. One also requires $q'(t_0)$ (the rate of increase of q at time t_0) to be reasonably large; it must at least be so large that over a period of time of length $(q'(t_0))^{-\frac{1}{2}}$ near to t_0 the changes in $L_1(t)$ and $L_2(t)$ are small, and $q'(t)$ itself must not appreciably alter in this period. Under these circumstances the integrand is negligible when w is considerably different from t_0, in comparison with its values at that time, and therefore one may replace $L_1(w)$ and $L_2(w)$ by $L_1(t_0)$ and $L_2(t_0)$, and $q'(w)$ by $q'(t_0)$. This gives the value

$$\sqrt{\pi}\,(q'(t_0))^{-\frac{1}{2}}[\beta_1(L_1(t_0))^2+\beta_2(L_2(t_0))^2]\exp\left[2\int_{t_0}^t q(z)\,dz\right], \tag{9.17}$$

for the variance of u.

The physical significance of this latter approximation is that the disturbances near the time when the instability is zero are the only ones which have any appreciable ultimate effect. Those which occur earlier are damped out by the subsequent period of stability. Those which occur later have a shorter period of instability within which to develop to greater amplitude. This principle is familiar in radio, and is fundamental to the theory of the superregenerative receiver.

Naturally one does not often wish to calculate the expression (9.17), but it is valuable as justifying a common-sense point of view of the matter. The factor $\exp\left[\int_{t_0}^t q(z)\,dz\right]$ is essentially the integrated instability and describes the extent to which one would expect disturbances of appropriate wave-length to grow between times t_0 and t. Taking the terms in β_1, β_2 into consideration separately, the factor $\sqrt{\pi}\beta_1(q'(t_0))^{-\frac{1}{2}}(L_1(t_0))^2$ indicates that the disturbances on the first morphogen should be regarded as lasting for a time

$$\sqrt{\pi}\,(q_1(t_0))^{-\frac{1}{2}}(bL_1(t_0))^2.$$

The dimensionless quantities $bL_1(t_0)$, $bL_2(t_0)$ will not usually be sufficiently large or small to justify their detailed calculation.

(3) The extent to which the component for which p_s is greatest may be expected to out-distance the others will now be considered in case (d). The greatest of the p_s will be called p_{s_0}. The two closest competitors to s_0 will be $s_0 - 1$ and $s_0 + 1$; it is required to determine how close the competition is. If the variation in the chemical data is sufficiently small it may be assumed that, although the exponents p_{s_0-1}, p_{s_0}, p_{s_0+1} may themselves vary appreciably in time, the differences $p_{s_0}-p_{s_0-1}$ and $p_{s_0}-p_{s_0+1}$ are constant. It certainly can happen that

one of these differences is zero or nearly zero, and there is then 'neck and neck' competition. The weakest competition occurs when $p_{s_0-1} = p_{s_0+1}$. In this case

$$p_{s_0} - p_{s_0-1} = p_{s_0} - p_{s_0+1} = -\tfrac{1}{2}(p_{s_0+1} - 2p_{s_0} + p_{s_0-1}).$$

But if s_0 is reasonably large $p_{s_0+1} - 2p_{s_0} + p_{s_0-1}$ can be set equal to $(\mathrm{d}^2p/\mathrm{d}s^2)_{s=s_0}$. It may be concluded that the rate at which the most quickly growing component grows cannot exceed the rate for its closest competitor by more than about $\tfrac{1}{2}(\mathrm{d}^2p/\mathrm{d}s^2)_{s=s_0}$. The formula (9·7), by which $\mathrm{d}^2p/\mathrm{d}s^2$ can be estimated, may be regarded as the product of two factors. The dimensionless factor never exceeds 4. The factor $\sqrt{(\mu'\nu')}/\rho^2$ may be described in very rough terms as 'the reciprocal of the time for the morphogens to diffuse a length equal to a radius'. In equally rough terms one may say that a time of this order of magnitude is required for the most quickly growing component to get a lead, amounting to a factor whose logarithm is of the order of unity, over its closest competitors, in the favourable case where $p_{s_0-1} = p_{s_0+1}$.

(4) Very little has yet been said about the effect of considering non-linear reaction rate functions when far from homogeneity. Any treatment so systematic as that given for the linear case seems to be out of the question. It is possible, however, to reach some qualitative conclusions about the effects of non-linear terms. Suppose that z_1 is the amplitude of the Fourier component which is most unstable (on a basis of the linear terms), and which may be supposed to have wave-length λ. The non-linear terms will cause components with wave-lengths $\tfrac{1}{2}\lambda, \tfrac{1}{3}\lambda, \tfrac{1}{4}\lambda, \ldots$ to appear as well as a space-independent component. If only quadratic terms are taken into account and if these are somewhat small, then the component of wave-length $\tfrac{1}{2}\lambda$ and the space-independent component will be the strongest. Suppose these have amplitudes z_2 and z_1. The state of the system is thus being described by the numbers z_0, z_1, z_2. In the absence of non-linear terms they would satisfy equations

$$\frac{\mathrm{d}z_0}{\mathrm{d}t} = p_0 z_0, \quad \frac{\mathrm{d}z_1}{\mathrm{d}t} = p_1 z_1, \quad \frac{\mathrm{d}z_2}{\mathrm{d}t} = p_2 z_2,$$

and if there is slight instability p_1 would be a small positive number, but p_0 and p_2 distinctly negative. The effect of the non-linear terms is to replace these equations by ones of the form

$$\frac{\mathrm{d}z_0}{\mathrm{d}t} = p_0 z_0 + A z_1^2 + B z_2^2,$$

$$\frac{\mathrm{d}z_1}{\mathrm{d}t} = p_1 z_1 + C z_2 z_1 + D z_0 z_1,$$

$$\frac{\mathrm{d}z_2}{\mathrm{d}t} = p_2 z_2 + E z_1^2 + F z_0 z_2.$$

As a first approximation one may put $\mathrm{d}z_0/\mathrm{d}t = \mathrm{d}z_2/\mathrm{d}t = 0$ and ignore z_1^4 and higher powers; z_0 and z_1 are then found to be proportional to z_1^2, and the equation for z_1 can be written $\mathrm{d}z_1/\mathrm{d}t = p_0 z_1 - k z_1^3$. The sign of k in this differential equation is of great importance. If it is positive, then the effect of the term $k z_1^3$ is to arrest the exponential growth of z_1 at the value $\sqrt{(p_1/k)}$. The 'instability' is then very confined in its effect, for the waves can only reach a finite amplitude, and this amplitude tends to zero as the instability (p_1) tends to zero. If, however, k is negative the growth becomes something even faster than exponential, and, if the equation $\mathrm{d}z_1/\mathrm{d}t = p_1 z_1 - k z_1^3$ held universally, it would result in the amplitude becoming infinite in a finite time. This phenomenon may be called 'catastrophic instability'. In the case of two-dimensional systems catastrophic instability is almost universal, and the corresponding equation takes the form $dz_1/dt = p_1 z_1 + k z_1^2$. Naturally enough in the case of catastrophic instability the amplitude does not really reach infinity, but when it is sufficiently large some effect previously ignored becomes large enough to halt the growth.

(5) Case (a) as described in § 8 represents a most extremely featureless form of pattern development. This may be remedied quite simply by making less drastic simplifying assumptions, so that a less gross account of the pattern can be given by the theory. It was assumed in § 9 that only the most unstable Fourier components would contribute appreciably to the pattern, though it was seen above (heading (3) of this section) that (in case (d)) this will only apply if the period of time involved is adequate to permit the morphogens, supposed for this purpose to be chemically inactive, to diffuse over the whole ring or organ concerned. The same may be shown to apply for case (a). If this assumption is dropped a much more interesting form of pattern can be accounted for. To do this it is necessary to consider not merely the components with $U = 0$ but some others with small positive values of U. One may assume the form $At - BU$ for p. Linearity in U is assumed because only small values of U are concerned, and the term At is included to represent the steady increase in instability. By measuring time from the moment of zero instability the necessity for a constant term is avoided. The formula (9·17) may be applied to estimate the statistical distribution of the amplitudes of the components. Only the factor $\exp\left[2\int_{t_0}^{t} q(z)\,dz\right]$ will depend very much on U, and taking $q(t) = p(t) = At - BU$, t_0 must be BU/A and the factor is

$$\exp\left[A(t - BU/A)^2\right].$$

The term in U^2 can be ignored if At^2 is fairly large, for then either $B^2 U^2/A^2$ is small or the factor e^{-BUt} is. But At^2 certainly is large if the factor e^{At^2}, applying when $U = 0$, is large. With this approximation the variance takes the form $C e^{-\frac{1}{2}k^2 U}$, with only the two parameters C, k to distinguish the pattern populations. By choosing appropriate units of concentration and length these pattern populations may all be reduced to a standard one, e.g. with $C = k = 1$. Random members of this population may be produced by considering any one of the type (a) systems to which the approximations used above apply. They are also produced, but with only a very small amplitude scale, if a homogeneous one-morphogen system undergoes random disturbances without diffusion for a period, and then diffusion without disturbance. This process is very convenient for computation, and can also be applied to two dimensions. Figure 2 shows such a pattern, obtained in a few hours by a manual computation.

To be more definite a set of numbers $u_{r,s}$ was chosen, each being ± 1, and taking the two values with equal probability. A function $f(x, y)$ is related to these numbers by the formula

$$f(x, y) = \Sigma u_{r,s} \exp\left[-\tfrac{1}{2}((x - hr)^2 + (y - hs)^2)\right].$$

In the actual computation a somewhat crude approximation to the function

$$\exp\left[-\tfrac{1}{2}(x^2 + y^2)\right]$$

was used and h was about $0\cdot7$. In the figure the set of points where $f(x, y)$ is positive is shown black. The outlines of the black patches are somewhat less irregular than they should be due to an inadequacy in the computation procedure.

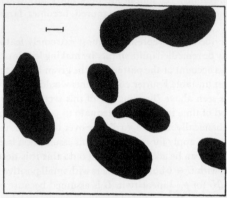

FIGURE 2. An example of a 'dappled' pattern as resulting from a type (a) morphogen system. A marker of unit length is shown. See text, §9, 11.

10. A NUMERICAL EXAMPLE

The numerous approximations and assumptions that have been made in the foregoing analysis may be rather confusing to many readers. In the present section it is proposed to consider in detail a single example of the case of most interest, (d). This will be made as specific as possible. It is unfortunately not possible to specify actual chemical reactions with the required properties, but it is thought that the reaction rates associated with the imagined reactions are not unreasonable.

The detail to be specified includes

  (i) The number and dimensions of the cells of the ring.

 (ii) The diffusibilities of the morphogens.

(iii) The reactions concerned.

(iv) The rates at which the reactions occur.

 (v) Information about random disturbances.

(vi) Information about the distribution, in space and time, of those morphogens which are of the nature of evocators.

These will be taken in order.

  (i) It will be assumed that there are twenty cells in the ring, and that they have a diameter of $0\cdot1$ mm each. These cells are certainly on the large rather than the small side, but by no means impossibly so. The number of cells in the ring has been chosen rather small in order that it should not be necessary to make the approximation of continuous tissue.

 (ii) Two morphogens are considered. They will be called X and Y, and the same letters will be used for their concentrations. This will not lead to any real confusion. The diffusion constant for X will be assumed to be $5 \times 10^{-8} \, \mathrm{cm^2 \, s^{-1}}$ and that for Y to be $2\cdot5 \times 10^{-8} \, \mathrm{cm^2 \, s^{-1}}$. With cells of diameter $0\cdot01$ cm this means that X flows between neighbouring cells at the rate 5×10^{-4} of the difference of X-content of the two cells per second. In other words, if there is nothing altering the concentrations but diffusion the difference of concentrations suffers an exponential decay with time constant 1000 s, or 'half-period' of 700 s. These times are doubled for Y.

If the cell membrane is regarded as the only obstacle to diffusion the permeability of the membranes to the morphogen is 5×10^{-6} cm/s or 0·018 cm/h. Values as large as 0·1 cm/h have been observed (Davson & Danielli 1943, figure 28).

(iii) The reactions are the most important part of the assumptions. Four substances A, X, Y, B are involved; these are isomeric, i.e. the molecules of the four substances are all rearrangements of the same atoms. Substances C, C', W will also be concerned. The thermo-dynamics of the problem will not be discussed except to say that it is contemplated that of the substances A, X, Y, B the one with the greatest free energy is A, and that with the least is B. Energy for the whole process is obtained by the degradation of A into B. The substance C is in effect a catalyst for the reaction $Y \rightarrow X$, and may also be regarded as an evocator, the system being unstable if there is a sufficient concentration of C.

The reactions postulated are

$$Y + X \rightarrow W,$$
$$W + A \rightarrow 2Y + B \quad \text{instantly,}$$
$$2X \rightarrow W,$$
$$A \rightarrow X,$$
$$Y \rightarrow B,$$
$$Y + C \rightarrow C' \quad \text{instantly,}$$
$$C' \rightarrow X + C.$$

(iv) For the purpose of stating the reaction rates special units will be introduced (for the purpose of this section only). They will be based on a period of 1000 s as units of time, and 10^{-11} mole/cm^3 as concentration unit*. There will be little occasion to use any but these special units (s.u.). The concentration of A will be assumed to have the large value of 1000 s.u. and the catalyst C, together with its combined form C' the concentration $10^{-3}(1 + \gamma)$ s.u., the dimensionless quantity γ being often supposed somewhat small, though values over as large a range as from -0.5 to 0.5 may be considered. The rates assumed will be

$$Y + X \rightarrow W \quad \text{at the rate } \tfrac{25}{16} YX,$$
$$2X \rightarrow W \quad \text{at the rate } \tfrac{7}{64} X^2,$$
$$A \rightarrow X \quad \text{at the rate } \tfrac{1}{16} \times 10^{-3} A,$$
$$C' \rightarrow X + C \quad \text{at the rate } \tfrac{55}{32} \times 10^{+3} C',$$
$$Y \rightarrow B \quad \text{at the rate } \tfrac{1}{16} Y.$$

With the values assumed for A and C' the net effect of these reactions is to convert X into Y at the rate $\tfrac{1}{32}[50XY + 7X^2 - 55(1 + \gamma)]$ at the same time producing X at the constant rate $\tfrac{1}{16}$, and destroying Y at the rate $Y/16$. If, however, the concentration of Y is zero and the rate of increase of Y required by these formulae is negative, the rate of conversion of Y into X is reduced sufficiently to permit Y to remain zero.

* A somewhat larger value of concentration unit (e.g. 10^{-9} mole/cm^3) is probably more suitable. The choice of unit only affects the calculations through the amplitude of the random disturbances.

In the special units $\mu = \frac{1}{2}$, $\nu = \frac{1}{4}$.

(v) Statistical theory describes in detail what irregularities arise from the molecular nature of matter. In a period in which, on the average, one should expect a reaction to occur between n pairs (or other combinations) of molecules, the actual number will differ from the mean by an amount whose mean square is also n, and is distributed according to the normal error law. Applying this to a reaction proceeding at a rate F (s.u.) and taking the volume of the cell as $10^{-8}\,\mathrm{cm}^3$ (assuming some elongation tangentially to the ring) it will be found that the root mean square irregularity of the quantity reacting in a period τ of time (s.u.) is $0 \cdot 004 \sqrt{(F\tau)}$.

TABLE 1. SOME STATIONARY-WAVE PATTERNS

cell number	first specimen				second specimen: incipient	'slow cooking': incipient	four-lobed equilibrium	
	incipient pattern		final pattern					
	X	Y	X	Y	Y	Y	X	Y
0	1·130	0·929	0·741	1·463	0·834	1·057	1·747	0·000
1	1·123	0·940	0·761	1·469	0·833	0·903	1·685	0·000
2	1·154	0·885	0·954	1·255	0·766	0·813	1·445	2·500
3	1·215	0·810	1·711	0·000	0·836	0·882	0·445	2·500
4	1·249	0·753	1·707	0·000	0·930	1·088	1·685	0·000
5	1·158	0·873	0·875	1·385	0·898	1·222	1·747	0·000
6	1·074	1·003	0·700	1·622	0·770	1·173	1·685	0·000
7	1·078	1·000	0·699	1·615	0·740	0·956	0·445	2·500
8	1·148	0·896	0·885	1·382	0·846	0·775	0·445	2·500
9	1·231	0·775	1·704	0·000	0·937	0·775	1·685	0·000
10	1·204	0·820	1·708	0·000	0·986	0·969	1·747	0·000
11	1·149	0·907	0·944	1·273	1·019	1·170	1·685	0·000
12	1·156	0·886	0·766	1·451	0·899	1·203	0·445	2·500
13	1·170	0·854	0·744	1·442	0·431	1·048	0·445	2·500
14	1·131	0·904	0·756	1·478	0·485	0·868	1·685	0·000
15	1·090	0·976	0·935	1·308	0·919	0·813	1·747	0·000
16	1·109	0·957	1·711	0·000	1·035	0·910	1·685	0·000
17	1·201	0·820	1·706	0·000	1·003	1·050	0·445	2·500
18	1·306	0·675	0·927	1·309	0·899	1·175	0·445	2·500
19	1·217	0·811	0·746	1·487	0·820	1·181	1·685	0·000

The diffusion of a morphogen from a cell to a neighbour may be treated as if the passage of a molecule from one cell to another were a monomolecular reaction; a molecule must be imagined to change its form slightly as it passes the cell wall. If the diffusion constant for a wall is μ, and quantities M_1, M_2 of the relevant morphogen lie on the two sides of it, the root-mean-square irregularity in the amount passing the wall in a period τ is

$$0 \cdot 004 \sqrt{\{(M_1 + M_2)\,\mu\tau\}}.$$

These two sources of irregularity are the most significant of those which arise from truly statistical cause, and are the only ones which are taken into account in the calculations whose results are given below. There may also be disturbances due to the presence of neighbouring anatomical structures, and other similar causes. These are of great importance, but of too great variety and complexity to be suitable for consideration here.

(vi) The only morphogen which is being treated as an evocator is C. Changes in the concentration of A might have similar effects, but the change would have to be rather great. It is preferable to assume that A is a 'fuel substance' (e.g. glucose) whose concentration does

[26]

not change. The concentration of C, together with its combined form C', will be supposed the same in all cells, but it changes with the passage of time. Two different varieties of the problem will be considered, with slightly different assumptions.

The results are shown in table 1. There are eight columns, each of which gives the concentration of a morphogen in each of the twenty cells; the circumstances to which these concentrations refer differ from column to column. The first five columns all refer to the same 'variety' of the imaginary organism, but there are two specimens shown. The specimens differ merely in the chance factors which were involved. With this variety the value of γ was allowed to increase at the rate of 2^{-7} s.u. from the value $-\frac{1}{4}$ to $+\frac{1}{16}$. At this point a pattern had definitely begun to appear, and was recorded. The parameter γ was then allowed to decrease at the same rate to zero and then remained there until there was no

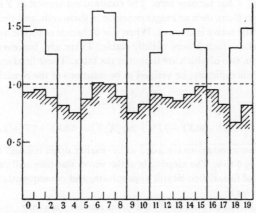

FIGURE 3. Concentrations of Y in the development of the first specimen (taken from table 1). ------- original homogeneous equilibrium; ////// incipient pattern; ——— final equilibrium.

more appreciable change. The pattern was then recorded again. The concentrations of Y in these two recordings are shown in figure 3 as well as in table 1. For the second specimen only one column of figures is given, viz. those for the Y morphogen in the incipient pattern. At this stage the X values are closely related to the Y values, as may be seen from the first specimen (or from theory). The final values can be made almost indistinguishable from those for the first specimen by renumbering the cells and have therefore not been given. These two specimens may be said to belong to the 'variety with quick cooking', because the instability is allowed to increase so quickly that the pattern appears relatively soon. The effect of this haste might be regarded as rather unsatisfactory, as the incipient pattern is very irregular. In both specimens the four-lobed component is present in considerable strength in the incipient pattern. It 'beats' with the three-lobed component producing considerable irregularity. The relative magnitudes of the three- and four-lobed components depend on chance and vary from specimen to specimen. The four-lobed component may often be the stronger, and may occasionally be so strong that the final pattern is four-lobed. How often this happens is not known, but the pattern, when it occurs, is shown in the last

8-2

[27]

two columns of the table. In this case the disturbances were supposed removed for some time before recording, so as to give a perfectly regular pattern.

The remaining column refers to a second variety, one with 'slow cooking'. In this the value of γ was allowed to increase only at the rate 10^{-5}. Its initial value was -0.010, but is of no significance. The final value was 0.003. With this pattern, when shown graphically, the irregularities are definitely perceptible, but are altogether overshadowed by the three-lobed component. The possibility of the ultimate pattern being four-lobed is not to be taken seriously with this variety.

The set of reactions chosen is such that the instability becomes 'catastrophic' when the second-order terms are taken into account, i.e. the growth of the waves tends to make the whole system more unstable than ever. This effect is finally halted when (in some cells) the concentration of Y has become zero. The constant conversion of Y into X through the agency of the catalyst C can then no longer continue in these cells, and the continued growth of the amplitude of the waves is arrested. When $\gamma = 0$ there is of course an equilibrium with $X = Y = 1$ in all cells, which is very slightly stable. There are, however, also other stable equilibria with $\gamma = 0$, two of which are shown in the table. These final equilibria may, with some trouble but little difficulty, be verified to be solutions of the equations (6·1) with

$$\frac{\mathrm{d}X}{\mathrm{d}t} = \frac{\mathrm{d}Y}{\mathrm{d}t} = 0,$$

and
$$32f(X, Y) = 57 - 50XY - 7Y^2, \quad 32g(X, Y) = 50XY + 7Y^2 - 2Y - 55.$$

The morphogen concentrations recorded at the earlier times connect more directly with the theory given in §§ 6 to 9. The amplitude of the waves was then still sufficiently small for the approximation of linearity to be still appropriate, and consequently the 'catastrophic' growth had not yet set in.

The functions $f(X, Y)$ and $g(X, Y)$ of § 6 depend also on γ and are

$$f(X, Y) = \tfrac{1}{32}[-7X^2 - 50XY + 57 + 55\gamma],$$
$$g(X, Y) = \tfrac{1}{32}[7X^2 + 50XY - 2Y - 55 - 55\gamma].$$

In applying the theory it will be as well to consider principally the behaviour of the system when γ remains permanently zero. Then for equilibrium $f(X, Y) = g(X, Y) = 0$ which means that $X = Y = 1$, i.e. $h = k = 1$. One also finds the following values for various quantities mentioned in §§ 6 to 9:

$$a = -2, \quad b = -1.5625, \quad c = 2, \quad d = 1.500, s = 3.333,$$
$$I = 0, \quad \alpha = 0.625, \quad \chi = 0.500, \quad (d-a)(-bc)^{-\frac{1}{2}} = 1.980,$$
$$(\mu+\nu)(\mu\nu)^{-\frac{1}{2}} = 2.121, \quad p_0 = -0.25 \pm 0.25\mathrm{i},$$
$$p_2 = -0.0648, \quad p_3 = -0.0034, \quad p_4 = -0.0118.$$

(The relation between p and U for these chemical data, and the values p_n, can be seen in figure 1, the values being so related as to make the curves apply to this example as well as that in § 8.) The value $s = 3.333$ leads one to expect a three-lobed pattern as the commonest, and this is confirmed by the values p_n. The four-lobed pattern is evidently the closest competitor. The closeness of the competition may be judged from the difference $p_3 - p_4 = 0.0084$, which suggests that the three-lobed component takes about 120 s.u. or about 33 h to gain an advantage of a neper (i.e. about $2\cdot7:1$) over the four-lobed one. However, the fact that γ is different from 0 and is changing invalidates this calculation to some extent.

The figures in table 1 were mainly obtained with the aid of the Manchester University Computer.

Although the above example is quite adequate to illustrate the mathematical principles involved it may be thought that the chemical reaction system is somewhat artificial. The following example is perhaps less so. The same 'special units' are used. The reactions assumed are

$$A \to X \qquad \text{at the rate} \quad 10^{-3}A, \ A = 10^3,$$
$$X + Y \to C \qquad \text{at the rate} \quad 10^3 XY,$$
$$C \to X + Y \qquad \text{at the rate} \quad 10^6 C,$$
$$C \to D \qquad \text{at the rate} \quad 62\cdot5 C,$$
$$B + C \to W \qquad \text{at the rate} \quad 0\cdot125 BC, \ B = 10^3,$$
$$W \to Y + C \qquad \text{instantly},$$
$$Y \to E \qquad \text{at the rate} \quad 0\cdot0625 Y,$$
$$Y + V \to V' \qquad \text{instantly},$$
$$V' \to E + V \qquad \text{at the rate} \quad 62\cdot5 V', \ V' = 10^{-3}\beta.$$

The effect of the reactions $X + Y \rightleftharpoons C$ is that $C = 10^{-3}XY$. The reaction $C \to D$ destroys C, and therefore in effect both X and Y, at the rate $\frac{1}{16}XY$. The reaction $A \to X$ forms X at the constant rate 1, and the pair $Y + V \to V' \to E + V$ destroys Y at the constant rate $\frac{1}{16}\beta$. The pair $B + C \to W \to Y + C$ forms Y at the rate $\frac{1}{8}XY$, and $Y \to E$ destroys it at the rate $\frac{1}{16}Y$. The total effect therefore is that X is produced at the rate $f(X, Y) = \frac{1}{16}(16 - XY)$, and Y at the rate $g(X, Y) = \frac{1}{16}(XY - Y - \beta)$. However, $g(X, Y) = 0$ if $Y \leqslant 0$. The diffusion constants will be supposed to be $\mu = \frac{1}{4}$, $\nu = \frac{1}{16}$. The homogeneity condition gives $hk = 16$, $k = 16 - \beta$.

It will be seen from conditions $(9\cdot4a)$ that case (d) applies if and only if $\frac{4}{k} + \frac{k}{4} < 2\cdot75$, i.e. if k lies between $1\cdot725$ and $9\cdot257$. Condition $(9\cdot4b)$ shows that there will be instability if in addition $\frac{8}{k} + \frac{k}{8} > \sqrt{3} + \frac{1}{2}$, i.e. if k does not lie between $4\cdot98$ and $12\cdot8$. It will also be found that the wave-length corresponding to $k = 4\cdot98$ is $4\cdot86$ cell diameters.

In the case of a ring of six cells with $\beta = 12$ there is a stable equilibrium, as shown in table 2.

TABLE 2

cell	0	1	2	3	4	5
X	7·5	3·5	2·5	2·5	3·5	7·5
Y	0	8	8	8	8	0

It should be recognized that these equilibria are only dynamic equilibria. The molecules which together make up the chemical waves are continually changing, though their concentrations in any particular cell are only undergoing small statistical fluctuations. Moreover,

8-3

in order to maintain the wave pattern a continual supply of free energy is required. It is clear that this must be so since there is a continual degradation of energy through diffusion. This energy is supplied through the 'fuel substances' $(A, B$ in the last example), which are degraded into 'waste products' (D, E).

11. Restatement and biological interpretation of the results

Certain readers may have preferred to omit the detailed mathematical treatment of §§ 6 to 10. For their benefit the assumptions and results will be briefly summarized, with some change of emphasis.

The system considered was either a ring of cells each in contact with its neighbours, or a continuous ring of tissue. The effects are extremely similar in the two cases. For the purposes of this summary it is not necessary to distinguish between them. A system with two or three morphogens only was considered, but the results apply quite generally. The system was supposed to be initially in a stable homogeneous condition, but disturbed slightly from this state by some influences unspecified, such as Brownian movement or the effects of neighbouring structures or slight irregularities of form. It was supposed also that slow changes are taking place in the reaction rates (or, possibly, the diffusibilities) of the two or three morphogens under consideration. These might, for instance, be due to changes of concentration of other morphogens acting in the role of catalyst or of fuel supply, or to a concurrent growth of the cells, or a change of temperature. Such changes are supposed ultimately to bring the system out of the stable state. The phenomena when the system is just unstable were the particular subject of the inquiry. In order to make the problem mathematically tractable it was necessary to assume that the system never deviated very far from the original homogeneous condition. This assumption was called the 'linearity assumption' because it permitted the replacement of the general reaction rate functions by linear ones. This linearity assumption is a serious one. Its justification lies in the fact that the patterns produced in the early stages when it is valid may be expected to have strong qualitative similarity to those prevailing in the later stages when it is not. Other, less important, assumptions were also made at the beginning of the mathematical theory, but the detailed effects of these were mostly considered in § 9, and were qualitatively unimportant.

The conclusions reached were as follows. After the lapse of a certain period of time from the beginning of instability, a pattern of morphogen concentrations appears which can best be described in terms of 'waves'. There are six types of possibility which may arise.

(a) The equilibrium concentrations and reaction rates may become such that there would be instability for an isolated cell with the same content as any one of the cells of the ring. If that cell drifts away from the equilibrium position, like an upright stick falling over, then, in the ring, each cell may be expected to do likewise. In neighbouring cells the drift may be expected to be in the same direction, but in distant cells, e.g. at opposite ends of a diameter there is no reason to expect this to be so.

This is the least interesting of the cases. It is possible, however, that it might account for 'dappled' colour patterns, and an example of a pattern in two dimensions produced by this type of process is shown in figure 2 for comparison with 'dappling'. If dappled patterns are to be explained in this way they must be laid down in a latent form when the foetus is only a few inches long. Later the distances would be greater than the morphogens could travel by diffusion.

(b) This case is similar to (a), except that the departure from equilibrium is not a uni-directional drift, but is oscillatory. As in case (a) there may not be agreement between the contents of cells at great distances.

There are probably many biological examples of this metabolic oscillation, but no really satisfactory one is known to the author.

(c) There may be a drift from equilibrium, which is in opposite directions in contiguous cells.

No biological examples of this are known.

(d) There is a stationary wave pattern on the ring, with no time variation, apart from a slow increase in amplitude, i.e. the pattern is slowly becoming more marked. In the case of a ring of continuous tissue the pattern is sinusoidal, i.e. the concentration of one of the morphogens plotted against position on the ring is a sine curve. The peaks of the waves will be uniformly spaced round the ring. The number of such peaks can be obtained approximately by dividing the so-called 'chemical wave-length' of the system into the circumference of the ring. The chemical wave-length is given for the case of two morphogens by the formula (9·3). This formula for the number of peaks of course does not give a whole number, but the actual number of peaks will always be one of the two whole numbers nearest to it, and will usually be *the* nearest. The degree of instability is also shown in (9·3).

The mathematical conditions under which this case applies are given in equations (9·4a), (9·4b).

Biological examples of this case are discussed at some length below.

(e) For a two-morphogen system only the alternatives (a) to (d) are possible, but with three or more morphogens it is possible to have travelling waves. With a ring there would be two sets of waves, one travelling clockwise and the other anticlockwise. There is a natural chemical wave-length and wave frequency in this case as well as a wave-length; no attempt was made to develop formulae for these.

In looking for biological examples of this there is no need to consider only rings. The waves could arise in a tissue of any anatomical form. It is important to know what wave-lengths, velocities and frequencies would be consistent with the theory. These quantities are determined by the rates at which the reactions occur (more accurately by the 'marginal reaction rates', which have the dimensions of the reciprocal of a time), and the diffusibilities of the morphogens. The possible range of values of the reaction rates is so immensely wide that they do not even give an indication of orders of magnitude. The diffusibilities are more helpful. If one were to assume that all the *dimensionless* parameters in a system of travelling waves were the same as in the example given in § 8, one could say that the product of the velocity and wave-length of the waves was 3π times the diffusibility of the most diffusible morphogen. But this assumption is certainly false, and it is by no means obvious what is the true range of possible values for the numerical constant (here 3π). The movements of the tail of a spermatozoon suggest themselves as an example of these travelling waves. That the waves are within one cell is no real difficulty. However, the speed of propagation seems to be somewhat greater than can be accounted for except with a rather large numerical constant.

(f) Metabolic oscillation with neighbouring cells in opposite phases. No biological examples of this are known to the author.

It is difficult also to find cases to which case (d) applies directly, but this is simply because isolated rings of tissue are very rare. On the other hand, systems that have the same kind of symmetry as a ring are extremely common, and it is to be expected that under appropriate chemical conditions, stationary waves may develop on these bodies, and that their circular symmetry will be replaced by a polygonal symmetry. Thus, for instance, a plant shoot may at one time have circular symmetry, i.e. appear essentially the same when rotated through any angle about a certain axis; this shoot may later develop a whorl of leaves, and then it will only suffer rotation through the angle which separates the leaves, or any multiple of it. This same example demonstrates the complexity of the situation when more than one dimension is involved. The leaves on the shoots may not appear in whorls, but be imbricated. This possibility is also capable of mathematical analysis, and will be considered in detail in a later paper. The cases which appear to the writer to come closest biologically to the 'isolated ring of cells' are the tentacles of (e.g.) *Hydra*, and the whorls of leaves of certain plants such as Woodruff (*Asperula odorata*).

Hydra is something like a sea-anemone but lives in fresh water and has from about five to ten tentacles. A part of a *Hydra* cut off from the rest will rearrange itself so as to form a complete new organism. At one stage of this proceeding the organism has reached the form of a tube open at the head end and closed at the other end. The external diameter is somewhat greater at the head end than over the rest of the tube. The whole still has circular symmetry. At a somewhat later stage the symmetry has gone to the extent that an appropriate stain will bring out a number of patches on the widened head end. These patches arise at the points where the tentacles are subsequently to appear (Child 1941, p. 101 and figure 30). According to morphogen theory it is natural to suppose that reactions, similar to those which were considered in connection with the ring of tissue, take place in the widened head end, leading to a similar breakdown of symmetry. The situation is more complicated than the case of the thin isolated ring, for the portion of the *Hydra* concerned is neither isolated nor very thin. It is not unreasonable to suppose that this head region is the only one in which the chemical conditions are such as to give instability. But substances produced in this region are still free to diffuse through the surrounding region of lesser activity. There is no great difficulty in extending the mathematics to cover this point in particular cases. But if the active region is too wide the system no longer approximates the behaviour of a thin ring and one can no longer expect the tentacles to form a single whorl. This also cannot be considered in detail in the present paper.

In the case of woodruff the leaves appear in whorls on the stem, the number of leaves in a whorl varying considerably, sometimes being as few as five or as many as nine. The numbers in consecutive whorls on the same stem are often equal, but by no means invariably. It is to be presumed that the whorls originate in rings of active tissue in the meristematic area, and that the rings arise at sufficiently great distance to have little influence on one another. The number of leaves in the whorl will presumably be obtainable by the rule given above, viz. by dividing the chemical wave-length into the circumference, though both these quantities will have to be given some new interpretation more appropriate to woodruff than to the ring. Another important example of a structure with polygonal symmetry is provided by young root fibres just breaking out from the parent root. Initially these are almost homogeneous in cross-section, but eventually a ring of fairly evenly spaced spots appear, and these later develop into vascular strands. In this case again the full explanation must be in terms of a two-dimensional or even a three-dimensional problem, although the analysis for the ring is still illuminating. When the cross-section is very large the strands may be in more than one ring, or more or less randomly or hexagonally arranged. The two-dimensional theory (not expounded here) also goes a long way to explain this.

Flowers might appear superficially to provide the most obvious examples of polygonal symmetry, and it is probable that there are many species for which this 'waves round a ring' theory is essentially correct. But it is certain that it does not apply for all species. If it did it would follow that, taking flowers as a whole, i.e. mixing up all species, there would be no very markedly preferred petal (or corolla, segment, stamen, etc.) numbers. For when all species are taken into account one must expect that the diameters of the rings concerned will take on nearly all values within a considerable range, and that neighbouring diameters will be almost equally common. There may also be some variation in chemical wave-length. Neighbouring values of the ratio circumferences to wave-length should therefore be more or less equally frequent, and this must mean that neighbouring petal numbers will have much the same frequency. But this is not borne out by the facts. The number five is extremely common, and the number seven rather rare. Such facts are, in the author's opinion, capable of explanation on the basis of morphogen theory, and are closely connected with the theory of phyllotaxis. They cannot be considered in detail here.

The case of a filament of tissue calls for some comment. The equilibrium patterns on such a filament will be the same as on a ring, which has been cut at a point where the concentrations of the morphogens are a maximum or a minimum. This could account for the segmentation of such filaments. It should be noticed, however, that the theory will not apply unmodified for filaments immersed in water.

12. CHEMICAL WAVES ON SPHERES. GASTRULATION

The treatment of homogeneity breakdown on the surface of a sphere is not much more difficult than in the case of the ring. The theory of spherical harmonics, on which it is based, is not, however, known to many that are not mathematical specialists. Although the essential properties of spherical harmonics that are used are stated below, many readers will prefer to proceed directly to the last paragraph of this section.

The anatomical structure concerned in this problem is a hollow sphere of continuous tissue such as a blastula. It is supposed sufficiently thin that one can treat it as a 'spherical shell'. This latter assumption is merely for the purpose of mathematical simplification; the results are almost exactly similar if it is omitted. As in §7 there are to be two morphogens, and a, b, c, d, μ', ν', h, k are also to have the same meaning as they did there. The operator ∇^2 will be used here to mean the superficial part of the Laplacian, i.e. $\nabla^2 V$ will be an abbreviation of

$$\frac{1}{\rho^2}\frac{\partial^2 V}{\partial \phi^2} + \frac{1}{\rho^2 \sin^2 \theta}\frac{\partial}{\partial \theta}\left(\sin\theta\frac{\partial V}{\partial \theta}\right),$$

where θ and ϕ are spherical polar co-ordinates on the surface of the sphere and ρ is its radius. The equations corresponding to (7·1) may then be written

$$\begin{aligned}
\frac{\partial X}{\partial t} &= a(X-h)+b(Y-k)+\mu'\nabla^2 X, \\
\frac{\partial Y}{\partial t} &= c(X-h)+d(Y-k)+\nu'\nabla^2 Y.
\end{aligned} \right\} \tag{12·1}$$

It is well known (e.g. Jeans 1927, chapter 8) that any function on the surface of the sphere, or at least any that is likely to arise in a physical problem, can be 'expanded in spherical surface harmonics'. This means that it can be expressed in the form

$$\sum_{n=0}^{\infty}\left[\sum_{m=-n}^{n} A_n^m P_n^m(\cos\theta)\,e^{im\phi}\right].$$

The expression in the square bracket is described as a 'surface harmonic of degree n'. Its nearest analogue in the ring theory is a Fourier component. The essential property of a spherical harmonic of degree n is that when the operator ∇^2 is applied to it the effect is the same as multiplication by $-n(n+1)/\rho^2$. In view of this fact it is evident that a solution of (12·1) is

$$\begin{aligned}
X &= h+\sum_{n=0}^{\infty}\sum_{m=-n}^{n}(A_n^m\,e^{iq_nt}+B_n^m\,e^{iq'_nt})\,P_n^m(\cos\theta)\,e^{im\phi}, \\
Y &= k+\sum_{n=0}^{\infty}\sum_{m=-n}^{n}(C_n^m\,e^{iq_nt}+D_n^m\,e^{iq'_nt})\,P_n^m(\cos\theta)\,e^{i\phi},
\end{aligned} \right\} \tag{12·2}$$

where q_n and q'_n are the two roots of

$$\left(q-a+\frac{\mu'}{\rho^2}n(n+1)\right)\left(q-d+\frac{\nu'}{\rho^2}n(n+1)\right)=bc \tag{12·3}$$

and

$$A_n^m\left(q_n-a+\frac{\mu'}{\rho^2}n(n+1)\right)=bC_n^m,$$

$$B_n^m\left(q'_n-a+\frac{\mu'}{\rho^2}n(n+1)\right)=cD_n^m. \tag{12·4}$$

This is the most general solution, since the coefficients A_n^m and B_n^m can be chosen to give any required values of X, Y when $t=0$, except when (12·3) has two equal roots, in which case a treatment is required which is similar to that applied in similar circumstances in §7. The analogy with §7 throughout will indeed be obvious, though the summation with respect to m does not appear there. The meaning of this summation is that there are a number of different patterns with the same wave-length, which can be superposed with various amplitude factors. Then supposing that, as in §8, one particular wave-length predominates, (12·2) reduces to

$$\begin{aligned}
X-h &= e^{iq_{n_0}t}\sum_{m=-n_0}^{n_0} A_{n_0}^m P_{n_0}^m(\cos\theta)\,e^{im\phi}, \\
b(Y-k) &= \left(q_{n_0}-a+\frac{\mu'}{\rho^2}n(n+1)\right)(X-h).
\end{aligned} \right\} \tag{12·5}$$

In other words, the concentrations of the two morphogens are proportional, and both of them are surface harmonics of the same degree n_0, viz. that which makes the greater of the roots q_{n_0}, q'_{n_0} have the greatest value.

It is probable that the forms of various nearly spherical structures, such as radiolarian skeletons, are closely related to these spherical harmonic patterns. The most important application of the theory seems, however, to be to the gastrulation of a blastula. Suppose that the chemical data, including the chemical wave-length, remain constant as the radius of the blastula increases. To be quite specific suppose that

$$\mu' = 2, \quad \nu' = 1, \quad a = -4, \quad b = -8, \quad c = 4, \quad d = 7.$$

With these values the system is quite stable so long as the radius is less than about 2. Near this point, however, the harmonics of degree 1 begin to develop and a pattern of form (12·5) with $n_0 = 1$ makes its appearance. Making use of the facts that

$$P_1^0(\cos\theta) = \cos\theta, \quad P_1^1(\cos\theta) = P_1^{-1}(\cos\theta) = \sin\theta,$$

it is seen that $X - h$ is of the form

$$X - h = A\cos\theta + B\sin\theta\cos\phi + C\sin\theta\sin\phi, \tag{12·6}$$

which may also be interpreted as

$$X - h = A'\cos\theta', \tag{12·7}$$

where θ' is the angle which the radius θ, ϕ makes with the fixed direction having direction cosines proportional to B, C, A and $A' = \sqrt{(A^2 + B^2 + C^2)}$.

The outcome of the analysis therefore is quite simply this. Under certain not very restrictive conditions (which include a requirement that the sphere be relatively small but increasing in size) the pattern of the breakdown of homogeneity is axially symmetrical, not about the original axis of spherical polar co-ordinates, but about some new axis determined by the disturbing influences. The concentrations of the first morphogen are given by (12·7), where θ' is measured from this new axis; and $Y - k$ is proportional to $X - h$. Supposing that the first morphogen is, or encourages the production of, a growth hormone, one must expect the blastula to grow in an axially symmetric manner, but at a greater rate at one end of the axis than at the other This might under many circumstances lead to gastrulation, though the effects of such growth are not very easily determinable. They depend on the elastic properties of the tissue as well as on the growth rate at each point. This growth will certainly lead to a solid of revolution with a marked difference between the two poles, unless, in addition to the chemical instability, there is a mechanical instability causing the breakdown of axial symmetry. The direction of the axis of gastrulation will be quite random according to this theory. It may be that it is found experimentally that the axis is normally in some definite direction such as that of the animal pole. This is not essentially contradictory to the theory, for any small asymmetry of the zygote may be sufficient to provide the 'disturbance' which determines the axis.

13. NON-LINEAR THEORY. USE OF DIGITAL COMPUTERS

The 'wave' theory which has been developed here depends essentially on the assumption that the reaction rates are linear functions of the concentrations, an assumption which is justifiable in the case of a system just beginning to leave a homogeneous condition. Such systems certainly have a special interest as giving the first appearance of a pattern, but they are the exception rather than the rule. Most of an organism, most of the time, is developing from one pattern into another, rather than from homogeneity into a pattern. One would like to be able to follow this more general process mathematically also. The difficulties are, however, such that one cannot hope to have any very embracing *theory* of such processes, beyond the statement of the equations. It might be possible, however, to treat a few particular cases in detail with the aid of a digital computer. This method has the advantage that it is not so necessary to make simplifying assumptions as it is when doing a more theoretical type of analysis. It might even be possible to take the mechanical aspects of the problem into account as well as the chemical, when applying this type of method. The essential disadvantage of the method is that one only gets results for particular cases. But this disadvantage is probably of comparatively little importance. Even with the ring problem, considered in this paper, for which a reasonably complete mathematical analysis was possible, the computational treatment of a particular case was most illuminating. The morphogen theory of phyllotaxis, to be described, as already mentioned, in a later paper, will be covered by this computational method. Non-linear equations will be used.

It must be admitted that the biological examples which it has been possible to give in the present paper are very limited. This can be ascribed quite simply to the fact that biological phenomena are usually very complicated. Taking this in combination with the relatively elementary mathematics used in this paper one could hardly expect to find that many observed biological phenomena would be covered. It is thought, however, that the imaginary biological systems which have been treated, and the principles which have been discussed, should be of some help in interpreting real biological forms.

REFERENCES

Child, C. M. 1941 *Patterns and problems of development.* University of Chicago Press.

Davson, H. & Danielli, J. F. 1943 *The permeability of natural membranes.* Cambridge University Press.

Jeans, J. H. 1927 *The mathematical theory of elasticity and magnetism,* 5th ed. Cambridge University Press.

Michaelis, L. & Menten, M. L. 1913 Die Kinetik der Invertinwirkung. *Biochem. Z.* **49,** 333.

Thompson, Sir D'Arcy 1942 *On growth and form,* 2nd ed. Cambridge University Press.

Waddington, C. H. 1940 *Organisers and genes.* Cambridge University Press.

A DIFFUSION REACTION THEORY OF
MORPHOGENESIS IN PLANTS

(The mathematical author (A.T.) is the originator of the theory (which will appear shortly in the *Proceedings of the Royal Society*) and is responsible for the exactitude with which his ideas and the relevant physical and mathematical concepts are represented, while the botanical author (C.W.W.) has tried to give the theory an appropriate botanical setting.)

I. *Introduction*

In contemporary studies of morphogenesis in plants, attention is being centred more and more on growth and the genic control of metabolism. It is held that the inception of new organs and the differentiation of tissues result from the localised accumulation of gene-determined substances. If this "substance" view of morphogenesis is accepted, the primary problems are then seen to relate to the nature of the "morphogenetic substances" and the factors which determine their patternised distribution. It is here that the student of morphogenesis has encountered one of his greatest difficulties: it seems improbable that biochemical concepts alone will enable him to give an adequate account either of the inception of pattern or of the progressive organisation during development which characterises the species. That specific substances, e.g. auxin, are of great importance in morphogenesis is now accepted, but thus far no adequate biochemical theory of organisation has been advanced. Such knowledge as we have of metabolic changes in embryonic regions does little to explain the assumption of form, the differentiation of tissues, and the orderly development of the individual species, with all its distinctive features. And although there is evidence that many of these developments are gene-controlled, our knowledge of the actual mechanism involved is both slender and speculative. On analysis, and as a working hypothesis, it seems that this mechanism, in its most fundamental aspect, must be sought in the laws of physical chemistry as applied to the metabolic systems found in embryonic regions. In particular, we have to inquire if anything is known regarding the physical chemistry of organic reaction systems which would account for the inception of some of the characteristic patterns*in plants. In a contemporary paper, TURING (1952) has advanced a theory, based on a comprehensive mathematical study of diffusion reaction systems, which goes a considerable way towards providing an explanation of the inception of pattern in living organisms. This theory is considered in its general botanical application in the present paper.

One major result of the comparative morphological studies of the post-Darwinian period, and also of the contemporary period of renewed interest in morphogenesis, has been an appreciation of the fact that similar morphological and anatomical features may be found in organisms of quite distinct taxonomic affinity. These *homoplastic* developments, which have resulted from parallel or convergent evolution, have also been aptly described as constituting *homologies of organisation* and are of general occurrence in the Plant Kingdom. Indeed, the main formal and structural features in plants can be referred to a comparatively small number of kinds of pattern (see section II). This being so, the factors which determine these kinds of pattern, or homologies of organisation, should be ascertained and closely investigated—a view already expressed by Lang in 1915.

In each instance where the assumption of form, or the inception of pattern, is being considered, it is essential to have some leading idea, or system of ideas, that will serve as a working basis for investigations. The contemporary explanation of comparable or *homogenous* developments in related organisms is that there are genes, or groups of genes, which are common to the organisms, and that these control or determine the observed developments**. But where similar features are present in unrelated organisms, the comparable developments cannot be attributed to common groups of genes. In attempting to explain the phenomenon of homology of organisation two possibilities may be entertained:

(i) comparable morphological features appear because essentially *the same kind of process* is operating in each of the non-related organisms; or

(ii) that essentially different processes may, nevertheless, yield comparable morphological results.

---

* Although, as ARBER (1950) has pointed out, the term *form* in its full connotation deals comprehensively with the characteristic shape of an organism, or of its parts, the term *pattern* has also been used extensively in morphological studies in recent years—a use that is not recorded in the *Oxford Dictionary*. Nevertheless, the term, being virtually self-explanatory, is a convenient one, and, in addition, it carries the implication that, as in an artistic design, a morphological or structural development may be characterised by repetitive features. The term is also useful in specifying particular aspects of the organisation which become manifest during development.

** Genetical investigations (HARLAND 1936; DE BEER 1951) show that the situation may be considerably more complicated.

On grounds of probability, the first explanation seems preferable to the second, but, because of the very great diversity of living organisms the second cannot, and should not, be eliminated out of hand. Whether we are concerned with developments which are considered to be more or less directly gene controlled, or with homologies of organisation, in which the importance of specific genetic factors has yet to be ascertained, the visible phenomena of morphogenesis have their inception in biochemical and biophysical reaction systems. In view of the prevelance of homology of organisation in the Plant Kingdom, it is cogent to inquire if there are any reaction systems which, because of their nature, are likely to be, or will necessarily be, of general occurrence in living organisms. Or, in other words, is anything known of reaction systems, of the kind likely to occur in plants, which would account for the inception of certain kinds of pattern?

A single example of homology of organisation may be considered by way of indicating the need for a new approach to these problems. In all classes of plants, the root stele as seen in cross-section consists of radiating plates of xylem alternating with bays of phloem. This pattern has its inception at the root apex and is not determined by the presence of lateral members. In small roots the pattern is simple, the xylem typically consisting of one or two to four or five xylem plates: in large roots there may be ten to twenty radiating xylem plates, i.e. the pattern, though considerably more complex, is essentially a repetition of that seen in small roots. Now, as a fact, no generally accepted hypothesis relating to the inception of pattern in root steles has yet been advanced. Familiarity with root structure may perhaps engender the impression that we understand what we see when we examine a cross-section under the microscope; but, in fact, we have, thus far, very little knowledge of the factors which determine the characteristic differentiation of the tissues. Broadly speaking, the same is true of all the tissue systems in plants; and the same general observation could be made regarding our knowledge of morphogenesis. GOEBEL (1922) pointed to the repetitive occurrence of pattern during development and to the relative constancy of scale of the "units of pattern" at the time of their inception. THODAY (1939) has indicated how this conception could be used to account for the increasing structural complexity in roots of increasing size, i.e. as the stele enlarges, more units of pattern can be accommodated. If this be accepted, then the fundamental problem is to discover the factors which determine the unit of pattern.

Every anatomical and morphological development is the result of antecedent physiological processes, and several, or many, steps may be involved. Hence any adequate theory of the inception of pattern (as in the root

[39]

stele given above) must take account of the nature and properties of embryonic tissue and of the physical chemistry of reacting organic substances—the metabolites involved in growth and differentiation. Furthermore, if we assume, as a working hypothesis, that substance differences are involved in the differentiation of phloem and xylem, then the problem is to account for the characteristic localisation of specific substances, i.e. for the patterned distribution of metabolites that precedes the visible tissue pattern. As it seems to the botanical author, Turing's diffusion-reaction theory of morphogenesis provides a new approach and contributes materially to our understanding of the mechanism of morphogenesis and histogenesis, particularly in their more general aspects. In the present paper the theory is briefly outlined and discussed (see section III). In the following section an indication is given to the various kinds of pattern found in plants to the inception of some of which the theory may have a special application.

II. *General analysis of pattern in plants*

Although the range and diversity in form and structure in plants are impressive, the number of distinctive *kinds of pattern* is small; but each may be greatly varied in the matter of detail. The following general categories or kinds of pattern may be indicated:

(1) *Axiate development*, which normally follows the early establishment of polarity (with an attendant physiological and morphological distinction between base and apex), is general in all classes, from algae to flowering plants.

(2) *Concentric construction* is exemplified by the cortex and stele in shoots and roots, by secondary thickening in shoots, by the wall and archesporium in bryophyte capsules, and so on.

(3) *Radiate construction* is typically seen in root steles, in the shoot stele in *Lycopodium*, in the bicollateral vascular strands in dicotyledons, etc. The branch filaments in *Chara* and other algae, and lateral members (leaves, buds, etc.) in vascular plants may also be included in this category.

(4) *Mosaic construction* is suggested as a term to denote such patterns as the distribution of stomata, the arrangement of tracheides and parenchyma in certain steles, etc.

(5) *Allometric development*: different rates of growth in different directions during development result in orderly changes in shape; the types of construction set out in (1)–(3) above may thus be greatly diversified.

(6) *Specific localisations*: this term is intended to include specifically

localised features such as the conceptacles in brown algae, the sori of ferns, the ovules in seed plants, etc.

The separation of these several kinds of pattern is, of course, artificial. During the growth of a root, for example, axiate, concentric and radiate developments are proceeding more or less simultaneously. Moreover, in some instances, both (4) and (6) above may indicate the same kind of pattern. This brief analysis does show, however, that the seemingly infinite diversity of plant form and structure can be referred to a few basic kinds of pattern.

III. *Turing's diffusion-reaction theory of morphogenesis*

Turing's theory (1952) is based on a consideration of the diffusibilities and reaction rates of substances which may be involved in growth and morphogenesis. Considerable mathematical knowledge is essential to follow the theory in detail, but its main features can be indicated to, and appreciated by, the non-mathematical biologist without too much difficulty. The theory introduces no new hypotheses: on the contrary, it makes use of well-known laws of physical chemistry, and, as Turing has shown, these seem likely to be sufficient to account for many of the facts of morphogenesis. The underlying point of view, in fact, is closely akin to that expressed by D'Arcy Thompson in *On Growth and Form*. It will be appreciated that a theory of this kind, based essentially on the laws of physical chemistry, is just what is needed to account for the fact that certain organisational features are of such general occurrence in plants. An essential feature of the theory is that it explains the inception of the morphogenetic pattern as a whole; but it is not inconsistent with epigenetic development when other organs have already been formed. Lastly, it is compatible with the concepts of physiological genetics.

An indication of the theory may be given. Let us assume that two interacting, pattern forming substances, or morphogens, X and Y, are essential metabolites in a morphogenetic process. A third substance C, which is in the nature of an evocator and a catalyst is also involved, a pattern only appearing if its concentration is sufficiently great. It is necessary to assume

(i) that both X and Y are diffusible, and at different rates; and

(ii) that there are a number of reactions involving X, Y and the catalyst C: these reactions do not merely use up the substances X and Y, but also tend to produce them from other metabolic substances (which might be called "fuel substances") which are assumed to be abundantly present in the growing region, i.e. to some extent the morphogens are autocatalytic.

If a pattern is to be produced, there are a number of conditions relating the diffusibilities and marginal reaction rates. (By marginal reaction rate is meant the amount by which the reaction rate changes per unit change of concentration.) If we assume that the appropriate conditions are satisfied, and that the concentration of the catalyst-evocator is initially at a low value, but is slowly increasing, the phenomena observed will then be as follows:

(i) Initially there is a state of homogeneity: both X and Y are uniformly distributed, apart from some slight deviations due to Brownian movement and to chance fluctuations in the number of X and Y molecules that have reacted in the various possible ways in various regions.

(ii) The concentrations will vary slowly as the system adjusts itself to the changing evocator concentration.

(iii) This change will also result in the fluctuations of concentration smoothing themselves out more and more slowly, and eventually the point is reached where the system is unstable, i.e. the fluctuations no longer are smoothed out: they become cumulative, and even tend to become exaggerated with the passage of time.

(iv) At this stage the morphogen concentrations form a more or less irregular wave pattern. Later, however (for instance when in some places the concentration of one morphogen is practically zero), the progressive deepening of the waves is arrested. The pattern will then regularise itself, and will eventually reach an equilibrium which is almost perfectly symmetrical. The resulting pattern may be described as a *stationary wave*.

(v) Such a stationary wave in a biological situation would be the equivalent of the accumulation of effective concentrations of the morphogens in regularly situated positions, e.g. 3, 4, 5 or more evenly distributed loci on a one-dimensional system such as a circle.

Put quite simply, all this is as much as to say that *in an embryonic process, in which the metabolic substances are originally distributed in a homogeneous manner, a regular patternised distribution of specific metabolites may eventually result and a morphological or histological pattern become manifest*. This patternised distribution of specific metabolites (or morphogenetic substances)—the stationary waves of the mathematician—takes place in conformity with the laws of physical chemistry as applied to diffusion-reaction systems. Given a reaction system, such as that described above, a regular distribution of loci of special metabolites will follow. In short, we see that, whether the morphogenetic substances are, or are not, specifically gene-determined, the inception of pattern can be referred to a physical system. It can scarcely be doubted that such reaction systems must be of general occurrence in living organisms. Acceptance of this view affords a basis for understanding the prevalence of homologies

[[42]]

of organisation, and also the diversification of basic kinds of pattern under the impact of genetic factors. For, as we have seen, the patternised distribution of metabolites depends on the reaction system, and the latter in turn depends on the diffusibility and chemical reaction of the metabolites, some of which may be specifically gene-determined. In short, the theory helps us to understand how certain morphogenetic patterns are of such general occurrence in the Plant Kingdom, and that this is necessarily so because they are the expression of physical systems normal to growing regions.

It may be difficult for some readers to understand how, from the initial homogeneous distribution of metabolites in an embryonic region, there can be a drift into instability as described in the theory. But Turing has indicated how the instability may be "triggered off" by random disturbances. Turing has envisaged an idealised and simplified "model of the embryo". "The model takes two slightly different forms. In one of them, the cell theory is recognized but the cells are idealised into geometrical points. In the other, the matter of the organisms is imagined as continuously distributed. The cells are not, however, completely ignored, for various physical and physico-chemical characterisations for the matter as a whole are assumed to have values appropriate to the cellular matter." In these statements the botanical reader will see that both the classical view of de Bary—that the tissue mass as a whole determines differentiation and not the individual cells ("Die Pflanze bildet Zellen, nicht die Zelle bildet Pflanzen")—and of physiological genetics (in which the importance of gene-controlled substances, proceeding from individual cells, is emphasised), are appropriately represented. The following may be cited as an example of Turing's approach: "With either of the models one proceeds as with a physical theory and defines an entity called the 'state of the system'. One then describes how that state is to be determined from the state at a moment very shortly before. With either model, the description of the state consists of two parts, the mechanical and the chemical. The mechanical part of the state describes the positions, masses, velocities and elastic properties of the cells, and the forces between them. In the continuous form of the theory, essentially the same information is given in the form of the stress, velocity, density and elasticity of the matter. The chemical part of the state is given (in the cell form of the theory) as the chemical composition of each separate cell: the diffusibility of each substance between each two adjacent cells must also be given. In the continuous form of the theory the concentrations and diffusibilities of each substance have to be given at each point. In determining the changes of state one should take into account:

(i) The changes of position and velocity as given by Newton's laws of motion.

(ii) The stresses as given by the elasticities and motions, also taking into account the osmotic pressures as given from the chemical data.

(iii) The chemical reactions.

(iv) The diffusion of the chemical substances. The region in which this diffusion is possible is given from the mechanical data.''

It will readily be appreciated that the mathematical treatment of changes in the state of even an arbitrary and greatly simplified diffusion reaction system is unavoidably complex. For the student of morphogenesis, however, it is the possibility of the general result rather than the details of the process that is important. Indeed, until it is disproved, the theory may be tentatively accepted; for it is based on known laws of physical chemistry and on a mathematical analysis of processes which, according to the present state of knowledge, are assumed to be going on in living organisms. What is perhaps most surprising is that the tangible results of complex metabolic processes are seen to be relatively simple, i.e. the orderly and often symmetrical inception of organs and tissues.

The idea that diffusion reaction systems are present in all growing regions, indeed in all living matter, is not new: it is basic to all studies of metabolism. What is novel in Turing's theory is his demonstration that, under suitable conditions, many different diffusion reaction systems will eventually give rise to stationary waves; in fact, to a patternised distribution of metabolites. If we consider an undifferentiated embryonic region, such as the apex of a root, in which a symmetrical, radiate histological pattern develops, the applicability of the theory seems highly probable. In that diffusion reaction systems are present in all growing regions, it would appear probable that they are, in some way, involved in the inception of pattern. Not all aspects of pattern however, can, or need, be referred to the development of stationary waves—the major feature of Turing's theory as thus far developed (1952). The inception of polarity, i.e. of axiate development in an embryo, is no doubt due to a patternised distribution of metabolites, but it is a different kind of pattern to that which has its inception in stationary waves. The following may be indicated as examples of pattern in plants which may be explained in whole or in part by the theory as presently developed: phyllotactic systems; whorled branching in algae; the distribution of procambial strands in shoots; the radiate pattern in root steles and in lycopod shoots. Turing has also indicated how the dappled pattern in the skins of animals and gastrulation in the developing embryo can be explained by his theory. It is not claimed that the inception of all kinds of pattern is due to stationary waves, but all will be referable to some kind of reaction system. Here it should perhaps be noted that the theory is still in the initial stages of development; but already it can be seen that it has a wide application and that this seems likely to be extended.

In the general system of ideas incorporated in the theory, there are many points of special interest to the student of morphogenesis. Thus, with reference to the breakdown of symmetry and homogeneity, attention is directed to the importance of small random changes in the distribution of morphogenetic substances, i.e. irregularities and statistical fluctuations in the numbers of molecules taking part in the various reactions. The experimental evidence indicates that the determination of polarity in the fertilised ovum of *Fucus* may be due to such random changes, including factors in the environment. In the enclosed embryos of land plants, in which polarity is held to be determined soon after fertilisation, quite small gradient effects proceeding from the parent gametophyte tissue may be the means of initiating the breakdown of homogeneity and the establishment of polarity. Some deviations from symmetry or homogeneity in a reaction system may be of great importance in the process of differentiation; for the system may reach a state of instability in which the irregularities, or certain components of them, tend to grow. If this happens, a new and stable equilibrium is usually reached, and this may show a considerable departure from the original symmetry, or distribution of metabolites. In contiguous cells, which are initially metabolically identical, a drift from equilibrium may take place in opposite directions as a result of statistical fluctuations in the components of the reaction system, or of small changes induced by neighbouring cells. Changes of this kind could, for instance, account for the very different developments in two adjacent, equivalent embryonic cells—a histological phenomenon that has long puzzled the botanist.

Unless we adopt vitalistic and teleological conceptions of living organisms, or make extensive use of the plea that there are important physical laws as yet undiscovered relating to the activities of organic molecules, we must envisage a living organism as a special kind of system to which the general laws of physics and chemistry apply. And because of the prevalence of homologies of organisation, we may well suppose, as D'Arcy Thompson has done, that certain physical processes are of very general occurrence. No contemporary biologist will deny that diffusion and the reaction of metabolic substances are the common and basic processes in the growth of organisms. Furthermore, as Turing has demonstrated, in some diffusion reaction systems, under certain conditions, the localised accumulation of specific substances necessarily results. A metabolic basis for the inception of pattern has thus been provided. If it stands up to such tests as can be devised, the great value of the theory in the study of morphogenesis is apparent: it is just what the botanist has been seeking for a very long time. It would be unwise, however, to expect the new theory to explain all the aspects of morphogenesis. In the determination of form and structure in plants, many factors of different kinds are at work, and any adequate approach to the problem must essentially be a multi-aspect one.

IV. *Tests of the theory*

It may be that for some time to come the theory will be provisionally accepted, at least by some biologists, because it rests on a substantial mathematical and physical basis, rather than because supporting data have been obtained. Certainly, tests of its validity should be sought. Because of the complexity of all morphogenetic processes, relevant experimental data may be difficult to obtain, but the task should not be regarded as an impossible one. An evident primary test of the theory will consist in the closeness of its applicability to a wide range of biological materials. On these grounds, as we have seen, it is likely to be found valid, since it cannot be denied that some diffusion reaction systems do give rise to stationary waves, and that such reaction systems are likely to be found in living organisms. That, of course, is not proof that stationary waves do, in fact, constitute the basis of pattern in plants, but the probability that this is so is considerable. The conception is certainly not incompatible with the ideas and data of contemporary plant physiology.

Indications of the validity of the theory by the method of prediction have already been obtained by Turing, using the digital computer. In a numerical example, in which two morphogens are considered to be present in a ring of 20 cells, he has found that a three- to four-lobed pattern would result; and in other examples he has shown that a two-dimensional pattern, such as dappling, and gastrulation in a spherical body, do arise in specified diffusion reaction systems. Here again, it should be noted that these results, based on reactions which approximate to those considered to take place in living organisms, increase the probability that the theory is valid or adequate, but do not prove that the inception of pattern is due to the assumed reaction systems.

In conclusion, it should be noted that the conception of diffusion reaction systems as the basis for morphogenesis raises its own difficulties. Not the least of these lies in the complexity and multiplicity of the processes which have to be envisaged. Thus, in any embryonic region such as a shoot apex, many different reactions are going on either simultaneously or in close succession, while different reactions may be taking place in contiguous tissues; and all these, together with the other factors which affect morphogenesis, result in the orderly, harmonious and characteristic development of the individual plant. The cautious biologist may well ask if situations of such complexity can indeed be unravelled and comprehended. The student of morphogenesis, moreover, accustomed as he is to work with visible and tangible materials, may find that the contemplation of diffusion reaction systems as required by the theory takes him into regions of thought with which he is unfamiliar—the realm of the unpicturable. Still, as it seems, if we are to break new ground in the study of morphogenesis and get to the root of the matter—having scratched so long at the surface—we must bear in mind that, since physiological processes always precede the appearance of new organs and tissues, it is with these processes that we are primarily concerned. And that being so, the new approach along the lines indicated in the diffusion reaction theory should be considered on its merits.

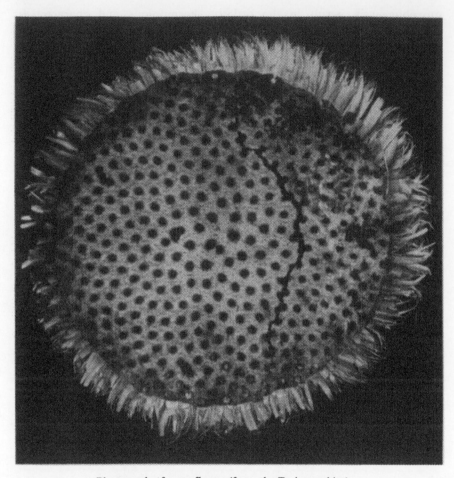

Photograph of a sunflower (from the Turing archive).

MORPHOGEN THEORY OF PHYLLOTAXIS

Part I. Geometrical and Descriptive Phyllotaxis

Phyllotaxis deals with the arrangements of leaves on the stems of plants. By a liberal interpretation of the terms "leaf" and "stem" it deals also with the arrangements of florets in a head (e.g. in a sunflower) and with the leaf primordia near the growing point of a bud. All these kinds of patterns will be discussed in the present paper. In the first part, which deals with some of the more superficial problems, the leaves are usually treated as if they were geometrical points distributed on a cylinder. Such patterns on cylinders are appropriate for the description of the mature structures, but their use may be criticised on the grounds that the patterns of real importance are not those formed by the mature structures, but of the leaf primordia. I would indeed go further and say that we should not consider even the primordia but certain patterns of concentration of chemical substances ("morphogens") which are present before there is any visible growth of primordia at all. This criticism is entirely valid, and the second part of the paper takes account of it. Nevertheless, a consideration of the patterns formed by the mature structures is enormously helpful, for a number of reasons.

(1) Suitable specimens of stems with leaves, large and robust enough for convenient examination, can be found almost anywhere, whereas the primordia can only be observed with the aid of a microscope and inconvenient techniques.

(2) The leaf patterns on a mature cylindrical stem are mathematically simpler and more easily intelligible than those near the growing point.

(3) In order to describe the patterns near the growing point satisfactorily it would in any case be necessary to carry through the mathematical theory of cylindrical patterns such as those formed by the mature structures, at least as an abstract discipline.

The method of exposition will be to alternate sections of mathematical theory with sections which describe facts about plants. The purpose of the theory may be lost if it is all given at once, and before any descriptive matter. On the other hand, if an attempt is made to describe first and theorise later the necessary terminology is lacking.

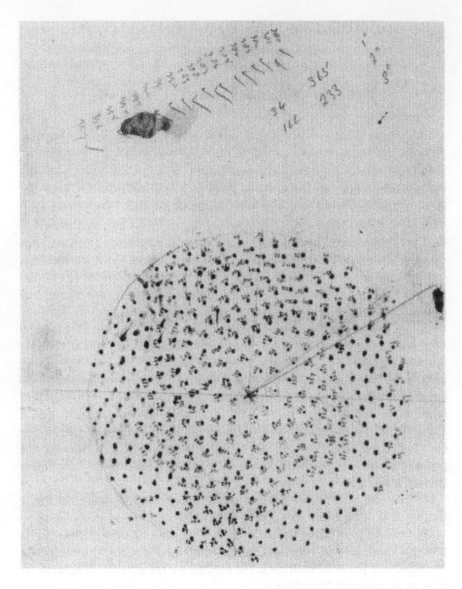

Diagram of a sunflower with the florets numbered (from the Turing archive).

1. *A description of certain leaf distribution patterns*

Plate 1 shows a portion of a branch of *Pinus* with a very regular arrangement of scales which at one time had supported leaves. The leaves have been removed to enable the patterns to be seen more clearly. The same pat-

Plate 1. A branch of *Pinus*. This photograph does not have the precise regularity described in the text (the original cannot be found) but the parastichies can be readily seen, especially where the leaves have been removed.

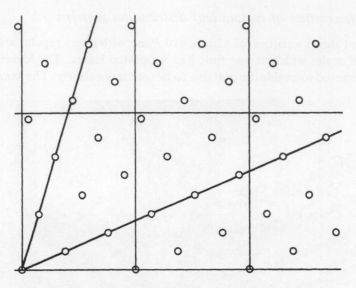

Fig. 1. An idealized plane representation of Plate 1.

tern is shown in Fig. 1 diagrammatically. The surface of the cylinder has here been unrolled onto the flat paper surface, and the whole enlarged. The scales have been reduced to points and will be referred to as "leaves". The three vertical lines represent one generator of the cylinder repeated. Each point between the first pair of lines represents a leaf which is also represented by one of the points between the second pair. The pattern is remarkably regular and is seen to have the following properties:

(1) If the cylinder is rotated and at the same time shifted along its length in such a way as to make a leaf A move into the position previously occupied by a leaf B, every other leaf also moves into a position previously occupied by a leaf. This may be called the *congruence property*.

(2) All the leaves lie at equal intervals along a helix. On the specimen in Plate 1 [i.e. the lost original], the pitch of the helix is about 0.046 cm and the successive leaves differ in angular position by about 137°.

These two properties are by no means independent. All patterns with the second property have also the first; but there are many species which produce leaf patterns having the first property but not the second. Figure 2 is a diagram similarly constructed to Fig. 1, and showing the arrangement of leaves on the stem of a maiden pink. In this arrangement each leaf has a partner at the same level with it on the stem. A helix could only pass through both partners if it had zero pitch and so degenerated into a circle. However, property (1) holds for these patterns also.

It need hardly be said that for the majority of botanical species the congruence property is only very roughly satisfied. But this need not trouble us for the present. It will be sufficient if the reader will admit that the congruence property has a certain botanical importance, and is willing in consequence to give some attention to the mathematics of patterns having the property.

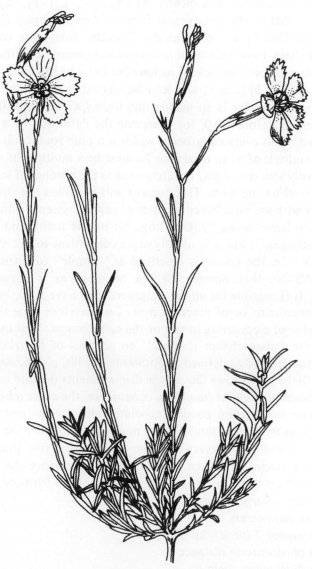

Fig. 2. A sketch of a maiden pink (*Dianthus deltoides*). Redrawn after Ross-Craig (1951).

2. *Helical coordinates for a phyllotactic system*

Consider the set of congruences of the patterns formed by the leaves on a stem, i.e. the set of pairs $[\theta, z]$ such that if the stem is simultaneously rotated through an angle θ about its axis, and shifted a distance z along it, both measured algebraically, each leaf is thereby moved into the position previously occupied by another leaf. If $[\theta_1, z_1]$ is one such congruence and $[\theta_2, z_2]$ is another, then clearly $[\theta_1 + \theta_2, z_1 + z_2]$, $[-\theta_1, -z_1]$ are also congruences, that is, the congruences form an Abelian group Γ. If n is an integer, then $[2n\pi, 0]$ is a congruence. Consider now those congruences which, like these, have a translation component (second coordinate) zero. The possible rotation components include 2π. Let κ be the smallest positive angle such that $[\kappa, 0]$ is in Γ, and let γ be any other such angle. One can write $\gamma = r\kappa + \delta$ where r is an integer and $0 \leqslant \delta \leqslant \kappa$. Then $[\delta, 0]$ is a congruence, and therefore $\delta = 0$, for otherwise the definition of κ would be contradicted. Thus every congruence which is a pure rotation is a rotation through a multiple of κ. In particular 2π must be a multiple of κ, $2\pi = J\kappa$, say. It is easily seen that J may be interpreted as the number of leaves which lie at one level on the stem. This number will be called the "jugacy", in conformity with the established practice of calling systems with $J = 2$ "bijugate" (two leaves being "yoked" together on the stem), and those with $J > 1$ "multijugate", but it is usually more convenient to use $\kappa = 2\pi/J$. If $J = 1$, i.e. $\kappa = 2\pi$, the system is described as "simple" (or in some books "alternate") but that phrase will not be used, as it suggests rather distichous. It is possible for all the congruences to have $z = 0$, but this case is too degenerate to be of much interest. Let therefore η be the smallest positive value of z occurring in any of the congruences. This quantity will be called the "plastochrone distance" on account of its relation to the "plastochrone ratio" as defined by RICHARDS (1948), p. 226. An argument similar to that above shows that all the displacements (second coordinates) are multiples of η. Now let $[\alpha, \eta]$ be a congruence, the angle α being chosen so as to have the smallest possible absolute value for the given η, and if this still leaves the sign in doubt, to be non-negative. Then $[n\alpha + r\kappa, n\eta]$ is a congruence, and indeed every congruence can be put into this form. For if $[\theta, n\eta]$ is a congruence, then so is $[\theta - n\alpha, 0]$; and since the translation component of the latter is zero, $\theta - n\alpha$ must be of the form $r\kappa$. The angle α is called the *divergence angle*.

The three parameters
- (i) the jugacy J (or $\kappa = 2\pi/J$),
- (ii) the plastochrone distance η,
- (iii) the divergence angle α

together completely describe the phyllotactic system, i.e. the total group of congruences. These three parameters, together with the radius of the cylinder, are the helical coordinates of the phyllotactic system.

3. *Parastichies and parastichy numbers*

In a diagram such as Fig. 1 showing the leaves on a stem, one can distinguish numerous straight lines with leaves at uniform intervals along them. These are known as *parastichies*. The word is commonly used for those series of leaves which most readily catch the eye, but no such restriction will be imposed in the present paper. A parastichy is thus the totality of leaves obtained by repeatedly applying the same congruence to some one leaf. Thus if a leaf has coordinates (θ_0, z_0) and $[\theta, z]$ is a congruence, then the leaves with coordinates $(\theta_0 + n\theta, z_0 + nz)$ form a parastichy. If one uses a different leaf, (θ_1, z_1) but the same congruence, one will in general obtain a different parastichy, running parallel to the first, though it may happen that one obtains the same one again. If the congruence $[\theta, z]$ is $[n\alpha + r\kappa, n\eta]$, then the cylinder includes $\eta^{-1} |n|^{-1}$ leaves of the parastichy per unit length. Since there are $J\eta^{-1}$ leaves per unit length altogether, there must be $|n| J = 2\pi |n| \kappa^{-1}$ parallel parastichies generated by the congruence $[n\alpha + r\kappa, n\eta]$. This explains the use of the term "parastichy number", for nJ is the number of different parastichies which the congruence generates, provided that $n > 0$. If $n = 0$, the parastichies are not helices on the cylinder but circles; each contains only a finite number of leaves, and there are infinitely many of them. It is preferable however to say in these cases that the parastichy number is zero rather than that it is infinite, so that the representation of the congruence as $[n\alpha + r\kappa, n\eta]$ may hold for all values of n, positive, zero or negative.

It is evident from the definition that if one adds two congruences, the parastichy number for the resulting congruence is obtained by adding the parastichy numbers for the two original congruences. This simple but important property is mentioned explicitly, since it is by no means so obvious when the parastichy numbers are defined by counting.

Note: What is here called "parastichy number" is called "leaf number difference" by botanists, whose own "parastichy" is a factor of our parastichy number. A parastichy with parastichy number 3 is indicated in Fig. 1.

4. *Phyllotactic systems as lattices. The principal congruences*

If ϱ is the radius of the cylinder, then $[\varrho\theta, z]$ will be called the *surface coordinates* of the congruence (or point) with cylindrical polar coordinates (θ, z). The surface coordinates of the congruences $[n\alpha + m\kappa, n\eta]$ of a phyllotactic system may be described as consisting of all the vectors $mu + nv$ where m, n are integers and u, v are respectively $(\varrho\kappa, 0)$ and $(\varrho\alpha, \eta)$. There are many other possible choices of u, v, e.g. $(3\varrho\alpha + 2\varrho\kappa, 3\eta)$ and $(\varrho\alpha + \varrho\kappa, \eta)$. The totality of vectors $mu + nv$ where m, n run over the integers and u, v are fixed vectors is called a *lattice*. In order that a lattice should arise from a phyllotactic system on a cylinder of radius ϱ it is necessary that $(2\pi\varrho, 0)$ should be a point of the lattice. This is also sufficient, as may be seen by interpreting a vector (y, x) of the lattice as a congruence $[y/\varrho, x]$.

One may define the *first principal vector* of a lattice as being that which is of shortest non-zero length. This defines it at best with a doubtful sign, and, as will appear later, at worst there are six equally valid candidates. It will be supposed that one of these is chosen to be the first principal vector; there is no need to enquire by what criteria. One may also define similarly the second, third principal vectors, etc. Each is to be the shortest consistent with not being a multiple of one of the earlier principal vectors.

In the sequel the first three principal vectors will play an important part. They correspond more or less to the "contact parastichies" of other investigators, the correspondence being closest for the parastichies generated by the first two principal vectors; but it has been thought that confusion would best be avoided by using an entirely different terminology.

It should be observed that the first two principal vectors of a lattice generate the lattice. For if not, a lattice-parallelogram must contain other lattice points within it. But a point within a parallelogram is always closer to one of the vertices than are some pair of the vertices from one another. Hence the definition of the principal vectors would be contradicted. A small consequence is that the first two principal vectors may also be defined as those two vertices which generate the lattice, and for which (subject to this condition) the square of the scalar product $(u \cdot v)$ has the minimum value.

Given two vectors u, v which generate the lattice, the value of $(u \cdot v)$ is increased by u^2 (or decreased by v^2) by replacing u by $u \pm v$ (or v by $v \pm u$). By repeatedly modifying the vectors in this way and reducing $|(u \cdot v)|$ without changing the sign of $(u \cdot v)$ one must eventually come to a pair for which $(u \cdot v)$ has the same sign as it had originally and $|(u \cdot v)| \leqslant u^2 \leqslant v^2$. By changing the sign of one of the original vectors if necessary, one may suppose this scalar product to be negative. Then

$$0 \leqslant -(u \cdot v) \leqslant u^2 \leqslant v^2$$

from which it follows that all three of the scalar products which can be formed from the vectors $u, v, -(u+v)$ are negative, i.e. the vectors form an acute angled triangle. If one had started with the principal vectors, no reduction would have been possible at all, so that the first three vectors must form an acute angled triangle. Conversely, *three vectors forming an acute angled triangle, any two of which generate the lattice, are the principal vectors.* For if u, v are the shortest and second shortest sides of the triangle respectively, then $|(u \cdot v)| \leqslant \frac{1}{2}u^2$, since v is shorter than $(u \pm v)$. Then if m, n are any two non-zero integers,

$$(mu + nv)^2 - v^2 = m^2u^2 + 2mn(u \cdot v) + (n^2 - 1)v^2$$
$$\geqslant (m^2 - |mn|)u^2 + (n^2 - 1)v^2$$
$$\geqslant (m^2 - |mn| + n^2 - 1)u^2 \quad \text{since } n^2 \geqslant 1 \text{ and } |v| \geqslant |u|$$
$$= ((|m| - |n|)^2 + |mn| - 1)u^2$$
$$\geqslant 0 \qquad\qquad\qquad \text{since } |m| \geqslant 1, |n| \geqslant 1.$$

Thus only vectors for which $m = 0$ or $n = 0$, i.e. only multiples of u or of v, can be shorter than v. Consequently u and v are the first two principal vectors.

These results may be summed up in the theorem on principal vectors:

The principal vectors form an acute angled triangle, and are the only vectors generating the lattice which do so. The first two principal vectors are also characterised by the property that they are the pair of vectors which generate the lattice and minimise the modulus of their scalar product.

In a phyllotactic lattice one may speak of the first, second, etc. principal parastichies and parastichy numbers. One then has the following simple consequence of the fact that $u \pm v$ is the third principal vector:

Corollary. *The third principal parastichy number is the sum or difference of the first and second parastichy numbers.*

Although it is not intended to enter into the matter yet in any detail, it may be mentioned that for a very large proportion of those plants which show sufficient regularity for parastichies to be counted, the principal parastichy numbers are all numbers of the Fibonacci series, in which each number after the first two is the sum of its two predecessors: $0, 1, 1, 2, 3, 5, 8, 13, 21, 34, 55, 89, \ldots$.

Clearly, if (say) the first two principal parastichy numbers are consecutive members of the series, the third and fourth must be also.

5. *The measurement of the phyllotaxis parameters*

It was explained in §2 that a phyllotaxis scheme is described by the parameters $\kappa = 2\pi/J, \alpha, \eta, \varrho$. On almost any specimen it is as well to measure the radius ϱ directly. On specimens on which the leaves are not very closely packed the jugacy $J = 2\pi/\kappa$ may be determined by counting how many leaves there are at any level on the stem. With more closely packed leaves this is not feasible, and it is best to choose two vectors which generate the lattice. The jugacy may then be determined as the highest common factor of two corresponding parastichy numbers. On specimens such as the stem shown in Plate 1, it is convenient to measure the distance and the angle between two leaves which are at a considerable distance apart.

To complete the calculation one must find the parastichy number corresponding to the congruence chosen, and the number of complete revolutions of the helix, which must be added to the angle measured. The parastichy number is obtained conveniently not by a direct count, but by counting two of the principle parastichies and combining the results by the addition rule. The divergence angle in such a case is best obtained by first making a less accurate measurement based on leaves which are not so far apart.

On more closely packed specimens it is better to choose two congruences (preferably principal congruences which generate the whole lattice) say $[m\alpha + r\kappa, m\eta]$ and $[n\alpha + s\kappa, n\eta]$, and measure the angles ψ_1, ψ_2 which the corresponding parastichies make with the generators of the cylinder. Then the area of the parallelogram generated by the first two principal vectors is

$$\Delta = mn\eta^2 |\tau_2 - \tau_1|$$

where $\tau_1 = \tan \psi_1$ and $\tau_2 = \tan \psi_2$. This area is also $\kappa\varrho\eta$, and therefore, since $\eta > 0$,

$$\eta = \kappa\varrho/mn |\tau_2 - \tau_1|.$$

The angle α satisfies

$$m\alpha = m\eta\varrho^{-1}\tau_1 \bmod \kappa \tag{I.5.1}$$

$$n\alpha = n\eta\varrho^{-1}\tau_2 \bmod \kappa \tag{I.5.2}$$

and since m, n are co-prime, positive integers, k, l can be found satisfying $km - ln = 1$. Therefore,

$$\alpha \equiv (km\tau_1 - ln\tau_2)\eta\varrho^{-1} \mod \kappa$$

$$\equiv \pm\left(\frac{2\pi k}{n}\frac{-\tau_1}{\tau_2-\tau_1} + \frac{2\pi l}{m}\frac{\tau_2}{\tau_2-t_1}\right). \tag{I.5.3}$$

The following rule expresses this formula in a convenient form. *Choose two vectors which generate the lattice and whose parastichy helices turn in opposite directions. Calculate (or look up in Table 1) what would be the divergence angle if either one of these parastichies were an orthostichy, i.e. parallel to the axis of the cylinder. The correct divergence angle may be obtained as a weighted average of these two. Each is to be weighted in proportion to the modulus of the cotangent of the angle which the corresponding parastichy makes with the generators of the cylinder.*

These angles are given with greater accuracy than can be used, though perhaps not so much greater as might be supposed. Since the angles given for the higher parastichy numbers differ by less than a minute, and since the angles ψ_1, ψ_2 can be measured to a few degrees, the divergence angle can be determined in such cases to within a few seconds. It need hardly be said that the value so obtained is not accurately repeated from leaf to leaf, and may vary by a degree or more, and it is only the averages over a considerable number of plastochrones that behave consistently. This insensitivity of the divergence angle to errors in the angles ψ_1, ψ_2 may be expressed in the equation

$$\left|\frac{d\alpha}{d\xi}\right| = \frac{\kappa}{mn}\frac{1}{(\xi-1)^2}.$$

Table 1
Divergence angles

Fraction of 2π	Deg., min., sec.			Degrees
1/2	180°			
1/3	120°			
2/5	144°			
3/8	135°			
5/13	138°	27′	41.5″	138.46154
8/21	137°	8′	34.3″	137.14286
13/34	137°	38′	49.4″	137.64706
21/55	137°	27′	16.4″	137.45454
34/89	137°	31′	41.1″	137.52809
55/144	137°	30′	00.0″	137.50000
89/233	137°	30′	38.6″	137.51073
Limiting value	137°	30′	27.9″	137.50778

In the case of limiting divergence-angle phyllotaxis ($\xi = 0$) this has the value κ/mn.

6. *Phyllotaxis on surfaces of revolution*

The patterns of leaves so far considered have been on the surface of a cylinder, and remain essentially the same on parts of the cylinder far removed from one another. Although species can be found for which, for the stems, this is a good approximation, some broader point of view is necessary to deal with the majority of phyllotactic patterns.

In general one may suppose that the specimen is a solid of revolution on which the lateral organs, idealised into points, are distributed. A common case is a capitulum, e.g. a sunflower or a daisy. The "leaves" are then florets and the surface of revolution is a disc, or nearly so.

For our purposes the geometry of the surface of revolution can be conveniently described as follows. The position of a point on the surface is fixed by two coordinates (θ, z) as on the cylinder. The coordinate z is measured *along the surface* (and not, as might be considered most natural, parallel to the axis). The shape of the surface is determined by giving the radius ϱ for each z. On such a surface one may define a phyllotactic system in which the parameters vary continuously with the coordinate z. The jugacy J, being an integer, cannot of course be allowed to vary at all. But suppose that at each z a value is assigned for the plastochrone distance η and the divergence angle α, as well as for the radius ϱ: what would be the positions of the leaves that correspond to arbitrary values of the parameters?

A natural answer can be given to this question if the formula $(n\alpha + r\kappa, n\eta)$ is extended to non-integral values of n, which we rename u. The formula is replaced by the two differential equations:

$$\frac{dz(u)}{du} = \eta(z), \qquad \frac{d\theta(u)}{du} = \alpha(z), \qquad (1.6.1)$$

the positions of the leaves being given by $(\theta(u) + r\kappa, z(u))$ for integral values of u and r.

With these conventions, one may obtain values of $z(n)$ and $\theta(n)$ by measurement and could, in theory, infer values of $\eta(z), \alpha(z)$ by ordinary finite difference methods. In practice there will be such errors of measurement, and irregularities in the positions of the leaves, that the use of differentiation formulae involving high differences is inappropriate. The method which the author finds most convenient is to draw freehand the principal parastichies in the neighbourhood of the value of z in question, measure the angles ψ_1, ψ_2 which these curves make with a plane through the axis (i.e. in practice with the intersection of this plane with the surface), count the parastichy numbers, and apply the formula of §5.

According to the point of view of this section there is a complete phyllotactic system corresponding to each value of the parameter z, described by parameters α, η, ϱ varying continuously with z. It will be convenient to continue to speak of such systems as if they were given on a cylinder, although η, α are defined by (I.6.1); and to consider the phyllotactic system as the lattice of points $(n\alpha + r\kappa, n\eta)$.

Attention will be given later (§11) to phyllotactic systems varying with a parameter.

7. *The bracket and the fractional notations*

When describing a specimen one may not always wish to make sufficient measurements to give a complete description of the lattice at some level on the stem: an indication of the principal parastichy numbers would often be enough. For this purpose the notation of CHURCH (1904) is appropriate. He used such notations as $(8 + 13)$, which in this paper will be used to signify that the principal parastichy numbers are 8, 13 and 21.

Some latitude must be allowed when the third and fourth principal vectors are of nearly equal length, and the three numbers may consequently be the first, second and fourth parastichy numbers. This cannot have been Church's intention, for he believed that two of the principal parastichies are always at right angles, an assumption which is not always correct.

Another, less happy notation is the use of fractions of a revolution as measures of the divergence angle. The most satisfactory approximations are of course the continued fraction convergents, and these will normally be the ratio of two Fibonacci numbers. That such ratios were good approximations to the divergence angle was first observed by Schimper and Braun (BRAUN 1835), and was an important discovery. However the use of different fractions of this kind to distinguish phyllotactic systems must be deplored. For instance, in the case where the divergence angle has the limiting value $2\pi\omega^{-2} = 137° \, 30' \, 27.9''$, all of these ratios are good approximations to the divergence angle. What then is the significance of the choice of one rather than another? On the whole, the tendency seems to be to choose larger denominators for smaller plastochrone ratios, but no very definite rule seems to have been formulated. In cases where there is some real reason for regarding the divergence angle as a rational fraction of a revolution, the use of such fractions is admissible. Such cases arise with distichous ($\alpha = 180°$, $J = 1$) and decussate ($\alpha = 90°$, $J = 2$) systems, and in fact with all symmetrical systems. Another example is provided by the genus *Carex*, where the stem itself is a triangular prism, thus ensuring that the divergence angle does not, on average, wander far from 120°. There are likewise species with a pentagonal stem (e.g. Plumbago) where the angle may be supposed to be 144°.

8. *Naturally occurring phyllotactic patterns*

It is found that the numbers in the Fibonacci series 0, 1, 1, 2, 3, 5, 8, 13, ... are by far the commonest parastichy numbers and a phyllotactic system with these numbers is described as normal. In these cases the divergence angle α is in the region 135° to 140° and if the principal parastichy numbers are large, α is near to 137°30′28″. However this system is not universal and other types of phyllotaxis mentioned below may be encountered.

(a) There are cases where the Fibonacci series is to be replaced by the series 1, 3, 4, 7, 11, ... (the "anomalous" series of Church). In some species this series is fairly common, but in others it appears only in a small proportion of specimens. For these cases the divergence angle is found to be in the neighbourhood of 99°30′.

(b) There are species (e.g. *Dipsacus sylvestris*) for which the principal parastichy numbers are taken from the double Fibonacci series 0, 2, 2, 4, 6, 10, 16, 26, ..., and the divergence angle is half the normal, i.e. about 68°45′14″.

(c) Some species have mirror symmetry, and indeed, this is true in the majority of cases where there are opposite leaves (i.e. bijugate phyllotaxis, $J = 2$). Commonest among these are the "decussate" leaf patterns, i.e. those for which the divergence angle is 90°, and the first four principal parastichy numbers are 0, 2, 2, 4 (not necessarily respectively). There are also cases where these parastichy numbers are 0, 1, 1, 2; and, relatively rarely, such combinations as 0, 6, 6, 12. The latter occur in species where the parastichy numbers are very variable, e.g. one might find within the species 0, 6, 6, 12; 1, 6, 7, 13; 0, 7, 7, 14; 0, 5, 5, 10; each forming a considerable fraction of the whole.

It is the main purpose of the present paper to explain, in part at any rate, the phenomena described above. The explanations given will be at two levels. In this first part of the paper the arguments are entirely geometrical. The geometrical arguments do not exactly give a theory of the development of a phyllotactic pattern. It is merely shown that if the development satisfies certain not very artificial conditions, then when once a phyllotactic pattern has started it will develop into patterns of the kind observed. These geometrical arguments have been expounded by some previous writers, but often in a rather unsatisfactory form, and with the emphasis misplaced.

The writer has consequently considered it appropriate to give a new exposition of these arguments. This first part of the paper is not however entirely old material in a new form. In particular the use of the inverse lattice and other ideas related to Fourier analysis appears to be new. This first and purely geometrical part of the paper must however be considered as merely a preliminary to the second part, which expounds a chemical explanation of the same phenomena. The chemical theory will be much more complex than the geometrical theory, and, in effect, justifies the assumptions of the latter. Although it might have been possible to expound the chemical theory totally independently of the geometrical, it was not thought advisable to do so, because of the insight which the geometrical theory gives.

9. *Lattice parameters*

It has been seen that the two principal vectors (a_0, b_0) and (c_0, d_0) generate the lattice. If these vectors, so far not uniquely specified, are precisely defined, their four coordinates can be used as parameters for describing the lattice, as alternatives to the helical parameters $\alpha, \eta, \varrho, J (= 2\pi/\kappa)$. Owing to the close connection with the principal parastichy numbers, the new parameters will be found more useful in theories of the origin of phyllotaxis. In order to make the definitions unique it is necessary to specify the signs that are to be given to the two vectors, and the order in which they are to be taken. It is convenient to require that the *second* coordinate of each vector should be non-negative. In phyllotactic systems this second coordinate, being an integral multiple of η, must be either at least as large as η or zero. In the latter case the convention will be that the first coordinate shall be positive, but this is rarely used in what follows. The ordering is to be such that $a_0 d_0 - b_0 c_0 > 0$. (In view of $b_0 \geq 0$, $d_0 \geq 0$, the condition for this is that the first vector can be made parallel to the second by turning it to the left through an angle of less than 180°.)

The conditions on the four numbers a_0, b_0, c_0, d_0 are thus that for every pair of non-zero integers m, n

$$a_0^2 + b_0^2 \leqslant (ma_0 + nc_0)^2 + (mb_0 + nd_0)^2, \tag{I.9.1}$$

$$c_0^2 + d_0^2 \leqslant (ma_0 + nc_0)^2 + (mb_0 + nd_0)^2, \tag{I.9.2}$$

$$b_0 > 0 \quad \text{or} \quad b_0 = 0 \text{ and } a_0 > 0, \tag{I.9.3}$$

$$d_0 > 0 \quad \text{or} \quad d_0 = 0 \text{ and } c_0 > 0, \tag{I.9.4}$$

$$a_0 d_0 - b_0 c_0 > 0. \tag{I.9.5}$$

[63]

Since the matrix

$$\begin{pmatrix} a_0 & b_0 \\ c_0 & d_0 \end{pmatrix}$$

plays a prominent part, the numbers (a_0, b_0, c_0, d_0) are called the *principal matrix coordinates* of the lattice. They are unique so long as the third principal vector is longer than the second. A further set of parameters suggested by these, having considerable intuitive appeal, are:

$$\Delta = a_0 d_0 - b_0 c_0, \tag{I.9.6}$$

$$\zeta = \left(\frac{a_0^2 + b_0^2}{c_0^2 + d_0^2} \right)^{1/2}, \tag{I.9.7}$$

$$\phi = -\sin^{-1} \left(\frac{a_0 c_0 + b_0 d_0}{[(a_0^2 + b_0^2)(c_0^2 + d_0^2)]^{1/2}} \right), \tag{I.9.8}$$

$$\psi = \tan^{-1} \left(\frac{a_0 + c_0}{b_0 + d_0} \right). \tag{I.9.9}$$

The letter Δ has already been used in its present sense of denoting the area of the parallelogram generated by the first two principal vectors, i.e., the area occupied by each leaf, and it may accordingly be called the *leaf area*; ζ is the ratio of the first two vectors; and ϕ is the angle between them reduced by 90°. The parameter ψ describes the direction of the sum of the first two principal vectors. It follows from the theorem on principal vectors that ϕ lies between $-30°$ and $30°$, and ψ between $-90°$ and $90°$. In practice, in phyllotactic lattices, $|\psi|$ does not often exceed 30°, while ζ is usually close to 1.

A lattice can be described by any pair of vectors which generate it. If (a, b) and (c, d) are two such vectors, the matrix

$$\begin{pmatrix} a & b \\ c & d \end{pmatrix}$$

will be called a *matrix representation* of the lattice. A necessary and sufficient condition that two matrices should describe the same lattice is that one should be obtainable from the other by left multiplication with a matrix with integral coefficients and determinant ± 1. By the second part of the theorem on principal vectors the principal representation of a lattice can be recognised by the fact that the vectors (a_0, b_0) and (c_0, d_0) form two of the sides of an acute angled triangle. It must of course also satisfy the conditions $b_0 \geqslant 0$, $d_0 \geqslant 0$, $a_0 d_0 - b_0 c_0 \geqslant 0$. If mJ and nJ are two parastichy numbers and if the corresponding vectors generate the lattice, and the parastichies make angles ψ_1, ψ_2 with the generators of the cylinder, then the matrix

$$\frac{\varrho\kappa}{\tau_2 - \tau_1} \begin{pmatrix} -\tau_1/n & 1/n \\ -\tau_2/m & 1/m \end{pmatrix}$$

is one of the matrix representations of the lattice. Here $\kappa = 2\pi/J$, $\tau_1 = \tan \psi_1$, $\tau_2 = \tan \psi_2$, and (cf. I.5.1)

$$\Delta = \frac{(\varrho\kappa)^2}{mn \, |\tau_2 - \tau_1|}. \tag{I.9.10}$$

To convert any matrix coordinates (a, b, c, d) for a lattice into helical coordinates η, α, ϱ, J one proceeds as follows. The value of η is easily obtained as the highest common factor of b and d. It is not possible to find the value of J since the same lattice may be wrapped around cylinders of various radii. For the present we suppose it given. To obtain $2\pi\varrho$ one must find the vectors of the lattice which have their first coordinates zero. If $b = mn$ and $d = n\eta$ then these vectors are clearly multiples of $(na - mc, 0)$, i.e. $\varrho\kappa = |na - mc|$. To obtain α let $km - ln = 1$, then $\alpha = (ka - lc)/\varrho$ modulo 2π.

10. *Continued fraction properties*

The procedure by which any matrix description of a lattice may be made to yield the principal description was described in effect in §4. Suppose that the scalar product of two vectors is negative. Then one repeatedly adds one vector to another, and continues until the modulus of the scalar product can no longer be reduced. Suppose that k is the largest integer such that the scalar product $(ka + c, kb + d) \cdot (a, b)$ is negative. Then after adding the first row (a, b) of the matrix k times to the second, it will be necessary to interchange the two rows, if the first is always to be added to the second. The effect of the combined addition and interchange is expressed by the multiplication of the matrix

$$\begin{pmatrix} a & b \\ c & d \end{pmatrix} \quad \text{on the left by} \quad \begin{pmatrix} k & 1 \\ 1 & 0 \end{pmatrix}.$$

The reduction process as a whole is then expressed by left multiplication by a product of a number of such matrices $C_{k_0}, C_{k_1}, \ldots, C_{k_r}$, where C_k represents

$$\begin{pmatrix} k & 1 \\ 1 & 0 \end{pmatrix}.$$

In order finally to bring the matrix to the form agreed as standard it may be necessary to left-multiply by one of the matrices

$$\begin{pmatrix} 0 & \pm 1 \\ \pm 1 & 0 \end{pmatrix} \quad \text{or} \quad \begin{pmatrix} \pm 1 & 1 \\ 0 & \pm 1 \end{pmatrix}.$$

Every unimodular matrix of order 2 can be expressed as a product

$$\begin{pmatrix} \pm 1 & 0 \\ 0 & \pm 1 \end{pmatrix} C_{k_1} C_{k_2} \cdots C_{k_r}.$$

Products $C_{k_0} C_{k_1} \cdots C_{k_r}$ are very closely related to continued fractions. In fact it can be shown by a simple inductive argument that if

$$K_0 + \frac{1}{K_1} + \frac{1}{K_2} + \frac{1}{K_3} + \cdots \frac{1}{K_r} = \frac{p_r}{q_r}$$

is in its lowest terms, then

$$C_{k_0} C_{k_1} \cdots C_{k_r} = \begin{pmatrix} p_r & p_{r-1} \\ q_r & q_{r-1} \end{pmatrix}.$$

This shows in effect that

Every improper unimodular matrix may be expressed in the form

$$\begin{pmatrix} p_r & q_r \\ p_{r-1} & q_{r-1} \end{pmatrix}$$

where p_r/q_r, p_{r-1}/q_{r-1} are two consecutive convergents of the continued fraction of some number. If the unimodular matrix is obtained by reduction of a matrix representation of a lattice then the partial quotients are given in reverse order as the number of times one vector is to be subtracted from the other without interchange.

This result may be applied to the lattice as described by the helical coordinates α, η, ϱ, J. One representative matrix is

$$2\pi\varrho J^{-1} \begin{pmatrix} -1 & 0 \\ x & \sigma \end{pmatrix}$$

where $\sigma = \eta J/2\pi\varrho$, $x = \alpha J/2\pi$. The lattice vectors are

$$2\pi\varrho J^{-1}(p \quad q) \begin{pmatrix} -1 & 0 \\ x & \sigma \end{pmatrix}$$

where p, q are any integers. It will be convenient to represent this vector by the expression (p/q): this notation is intended to suggest a connection with fractions, but the brackets are always to be retained to prevent any confusion. By what has been shown above, a standard representative, apart from

[66]

the order of the rows, can be written in the form

$$2\pi\varrho J^{-1}\begin{pmatrix} q_n x - p_n & \sigma q_n \\ q_{n-1}x - p_{n-1} & \sigma q_{n-1} \end{pmatrix}.$$

The first two principal vectors are then, not necessarily in order,

$$u = (p_n/q_n), \qquad v = (p_{n-1}/q_{n-1}).$$

Denoting the third vector by (p/q), the theorem on principal vectors and acute angled triangles gives

$$(p/q) = (p_n \pm p_{n-1}/q_n \pm q_{n-1}).$$

Now the three quantities $q_n x - p_n$, $q_{n-1}x - p_{n-1}$, $qx - p$ cannot all have the same sign, for if they did the three scalar products (p_n/q_n), (p_{n-1}/q_{n-1}), (p/q) would all be positive, contrary to the results of the same theorem. If $q_n x - p_n$ and $q_{n-1}x - p_{n-1}$ have opposite signs, then p_{n-1}/q_{n-1} is a convergent of x. If, however, they have the same sign, (p/q) must be $(p_n - p_{n-1}/q_n - q_{n-1})$ and $|q_{n-1}x - p_{n-1}| > |q_n x - p_n|$. Since

$$p_n = K_n p_{n-1} + p_{n-2}, \qquad q_n = K_n q_{n-1} + q_{n-2}, \quad \text{and} \quad K_n \geqslant 1,$$

it follows that $q_{n-2}x - p_{n-2}$ also has the opposite sign to $q_{n-1}x - p_{n-1}$, and therefore p_{n-2}/q_{n-2} is a convergent of x. If $K_n = 1$ then (p_{n-2}/q_{n-2}) is the third vector. If, however, $K_n > 1$ then

$$|q_{n-2}x - p_{n-2}| = (K_n - 1)|q_{n-1}x - p_{n-1}| + |qx - p|$$

$$> |q_{n-1}x - p_{n-1}|. \tag{I.10.1}$$

Thus in any case one of the three principal vectors, that with the smallest or second smallest parastichy number, corresponds to a convergent of x.

Rather more may be asserted in the conditions which normally apply in real phyllotaxis. Let p_{r-1}/q_{r-1}, p_r/q_r, p_{r+1}/q_{r+1} now represent three consecutive convergents of $x = \alpha J/2\pi$, and suppose that

$$-\left(x - \frac{p_{r-1}}{q_{r-1}}\right)\left(x - \frac{p_r}{q_r}\right) > \sigma^2$$

$$> -\left(x - \frac{p_r}{q_r}\right)\left(x - \frac{p_{r+1}}{q_{r+1}}\right) > 0, \tag{I.10.2}$$

$$p_{r+1} = p_r + p_{r-1}, \qquad q_{r+1} = q_r + q_{r-1}, \tag{I.10.3}$$

then it follows at once that the scalar product $((p_{r-1}/q_{r-1})\cdot(p_r/q_r))$ is negative and that $((p_r/q_r)\cdot(p_{r+1}/q_{r+1}))$ and $((p_{r-1}/q_{r-1})\cdot(p_{r+1}/q_{r+1}))$ are positive. Also $(p_{r+1}/q_{r+1}) = (p_r/q_r) + (p_{r-1}/q_{r-1})$. Thus these three vectors form an acute angled triangle, and therefore are the three principal vectors of the lattice. Thus

If the divergence angle as a fraction of $2\pi/J$ has one of its partial quotients a_{r_0+1} equal to unity, so that

$$p_{r_0+1}=p_{r_0}+p_{r_0-1}, \qquad q_{r_0+1}=q_{r_0}+q_{r_0-1},$$

then for some values of the plastochrone ratio η/ϱ the principal vectors correspond to these three consecutive convergents of the divergence angle, though not necessarily in order. If there are a number of consecutive unit partial quotients, the corresponding ranges of values of the plastochrone ratio are consecutive intervals.

In particular, if all the partial quotients are unity from some point onwards, then for all sufficiently small plastochrone ratios the principal vectors all correspond to convergents of the divergence angle. Such a divergence angle may be called a *limiting divergence angle*. The reduced parastichy numbers from some point onwards are the denominators q_r and satisfy $q_{r+1}=q_r+q_{r-1}$. They determine the value of the limiting divergence angle, apart of course for sign and for additive multiples of 2π. For the numerators p_r must satisfy $p_{r+1}=p_r+p_{r-1}$ and $q_r p_{r-1}-q_{r-1}p_r=\pm1$. A change of sign in the value of $q_r p_{r-1}-q_{r-1}p_r$ may be accomplished by changing the sign of each p_r and therefore of the limiting angle

$$\lim_{r\to\infty}\frac{2\pi}{J}\frac{p_r}{q_r}.$$

Suppose then that

$$q_{r_0}p_{r_0+1}-q_{r_0+1}p_{r_0}=1$$

and that p_r' is a second solution so that

$$q_r(p_{r+1}-p_{r+1}')=q_{r-1}(p_r-p_r').$$

Then since q_r, q_{r+1} have no common factor, for some integer m one must have

$$p_{r+1}'=p_{r+1}+mq_{r+1}, \qquad p_r'=p_r+mq_r$$

and p_r/q_r differs from p_r'/q_r' only by the integer m. If p_r'/q_r' tends to a limit it can only differ from the limit of p_r/q_r by an integer, and hence from the corresponding estimates of α by a multiple of κ, which may be eliminated by the minimal condition on α. For this reason, the limiting angle may be described as "the limiting divergence angle corresponding to the series $q_{r_0-1}, q_{r_0}, q_{r_0+1}$ of reduced parastichy numbers".

[68]

The value of the limiting divergence angle may be expressed as

$$\kappa \frac{p_{r_0} + \omega p_{r_0+1}}{q_{r_0} + \omega q_{r_0+1}}.$$

The cases of chief interest are:

Normal Fibonacci phyllotaxis. The parastichy numbers belong to the series $0, 1, 1, 2, 3, 5, \ldots$ and the numerators to the same series displaced: $-1, 1, 0, 1, 1, 2, \ldots$. The limiting angle is $2\pi(1 - \omega^{-1})$, i.e. $137° \, 30' \, 28''$.

Normal bijugate phyllotaxis. The parastichy numbers are $0, 2, 2, 4, \ldots$, i.e. the Fibonacci numbers doubled. The jugacy is two, and the reduced parastichy numbers are again the Fibonacci numbers. The limiting angle is $\pi(1 - \omega^{-1})$ i.e. $68° \, 45' \, 14.0''$.

Phyllotaxis of the so-called anomalous series. The principle parastichy numbers belong to the series $1, 3, 4, 7, 11, \ldots$. The limiting angle is $2\pi\omega/(1 + 3\omega)$, i.e. $99° \, 30' \, 6''$.

Examples were also collected by the brothers BRAVAIS (1838) of the occurrence of other series.

11. *Continuously changing phyllotaxis*

There are at least two ways in which one may be concerned with a phyllotactic system which depends continuously on some real parameter. On a growing specimen the dimensions of the stem may be altering and the leaves moving relative to one another. In this way one is concerned with a phyllotaxis which depends on time. One may equally well be concerned with a phyllotaxis which changes continuously along the length of a stem, or towards the centre of a capitulum. Variation of phyllotaxis in time will be given greater attention here than variation in space, though it can be less conveniently demonstrated. For the purpose of the present section either kind of variation will be described by allowing α, η, ϱ and other quantities describing the phyllotaxis to depend upon a parameter t, which will be described as "the time" regardless of the fact that other interpretations would be equally appropriate.

Until comparatively recently, phyllotaxis continuously varying in space has only been considered in the rather trivial case in which although the radius ϱ is allowed to vary with position (as indeed on a disc it is obliged to do), the dimensionless quantities $\alpha, \eta/\varrho, J$ remain constant. In such a system all other dimensionless quantities, such as angles, principal parastichy numbers, etc., also remain constant, and it may be considered that one has effectively the same phyllotaxis at each radius. This hypothesis may be most familiar to the reader in the form in which it is stated that the parastichies are logarithmic spirals. It may be appropriate for certain conditions of exponential growth, but there are also many conditions for which it is entirely inappropriate. For instance in the case of the florets on a mature capitulum of *Taraxacum officinale* the area per floret, so far from varying inversely as the square of the radius, as would be required on this hypothesis, actually increases towards the centre. It is also inappropriate for the neighbourhood of the growing point, at least if there be any truth in the theory expounded in part II. It would be particularly inappropriate to restrict consideration of continuously varying phyllotaxis to this case, which does not admit of changes of principal parastichy numbers, since it is intended to explain in part these changes in terms of continuously changing phyllotaxis.

More recently RICHARDS (1948) has considered phyllotactic systems on a disc with the divergence angle always equal to the Fibonacci angle, i.e.

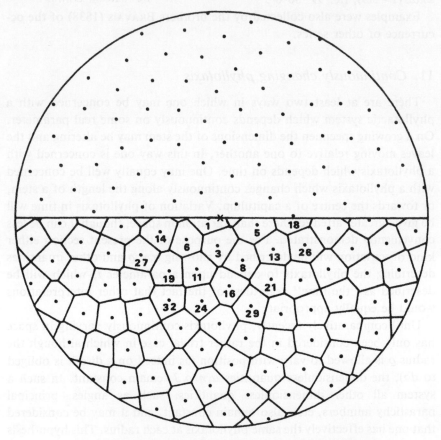

Fig. 3. A phyllotactic system on a disc. After RICHARDS (1948).

the limiting angle for the Fibonacci series, viz. $2\pi\omega^{-2}$, and the radius varying with the leaf number in various ways. In general if $f(u)$ is some increasing function of the real variable u one may consider leaves to be placed at the points described in plane polar coordinates by $(f(u), 2\pi\omega^{-2}u)$.

According to the principles of §6, one has $z = \varrho$ and therefore $z(u) = f(u)$. The value of η for radius ϱ is $f'(u)$ where $\varrho = f(u)$. A particularly interesting example is provided by putting $f(u) = A(u + B)^{1/2}$. Then $\eta = A^2/2\varrho$, and the leaf area has the same value πA^2 at each radius: this is quite a good approximation for the distribution of mature florets in the head of a member of the compositae. Figure 3 shows such a phyllotaxis. It is taken from RICHARDS (1948) and in it $B = 1/2$.

It follows from the results of the last section that the principal parastichy numbers for this phyllotactic pattern are Fibonacci numbers at each radius. One must be cautious here and avoid assuming causal connections groundlessly. Certainly if the divergence angle is exactly the Fibonacci angle then the principal parastichy numbers must be Fibonacci numbers, and conversely, if the principal parastichy numbers are Fibonacci numbers the divergence angle cannot differ much from the Fibonacci angle. If one of these phenomena has to be the cause of the other, then the less objectionable assumption is that the value of the angle is the effect. But there is no need to adopt either hypothesis. In view of the corollary to the theorem on principal parastichy numbers and acute angled triangles (§4) it is not possible for the first two of the principal parastichy numbers to be Fibonacci numbers without the third being one also. It is to be expected that this fact should go a long way to explain the great preponderance of Fibonacci numbers amongst the principal parastichy numbers. Clearly, however, some other hypothesis is necessary in addition, for it is possible to produce (as mathematical constructions) phyllotactic systems with any two given integers as the first two principal parastichy numbers. An appropriate hypothesis for the purpose is the following:

The third principal parastichy number does not lie numerically between the first and second.

This will be known as the *hypothesis of geometrical phyllotaxis.* It can easily be seen by examination of Fig. 3 that it applies for Fibonacci angle phyllotaxis, and it may also be shown to apply for any limiting angle phyllotaxis. It appears also to apply to all naturally occurring cases. (To verify it on a specimen, look for the acute angled triangles. The longest side should either join the uppermost and lowermost points of the triangle, or else be sufficiently nearly horizontal for its projection onto a vertical line to be less than half the projection of the whole triangle.) It is adequate as a subsidiary condition for Fibonacci phyllotaxis for

If a phyllotactic system varies in time whilst satisfying the hypothesis of geometrical phyllotaxis, then the three principal parastichy numbers of the system always belong to the same sequence p_r obeying the Fibonacci law $p_{r+1} = p_r + p_{r-1}$.

If this be not so then there must be a time when the principal parastichy numbers change from a set p_{r-1}, p_r, p_{r+1} obeying the rule to a set which do not. This must arise through the original third parastichy number being dropped out and being replaced by another number. It cannot be p_{r-1} which is dropped, for then p_r, p_{r+1} would remain, and by §4 the third parastichy number after the change must be either $p_{r+1} - p_r$ or $p_{r+1} + p_r$, i.e. either p_{r-1} or p_{r+2}, either of which would belong to the series contrary to hypothesis. Likewise it cannot be p_{r+1} which is dropped. It cannot be p_r for this would contradict the hypothesis of geometrical phyllotaxis. Thus each alternative is impossible, and the assumption that it is possible to reach principal parastichy numbers not in the series p_r is contradicted.

However, this hypothesis cannot be regarded as entirely satisfactory. However true it is, and however logically it follows that the principal parastichy numbers remain in a Fibonacci-like series, the hypothesis is itself quite arbitrary and unexplained. Its merit is that it replaces an empirical law, of a rather weird and magical appearance, by something simpler and much less mysterious. The question remains "Why should the hypothesis of geometrical phyllotaxis hold?" and this is a question which the geometrical approach is not capable of answering. There are other questions which it is also unable to answer, such as "Why should leaf patterns take the form of lattices at all?", "How does the lattice pattern develop on previously undifferentiated pieces of tissue?" and "How does a mirror symmetrical pattern develop into an unsymmetrical one?" and "Why are the principal vectors of the lattice of such importance?" All of these topics must be left until part II. There remain however a number of questions which can be treated by the methods of geometrical phyllotaxis, which have not yet been considered. Some of these are preparatory for the work of part II; others throw further light on the mathematics of continuously changing phyllotaxis and the mathematical description of phyllotactic lattices, and yet others are concerned with the packing of mature and semi-mature leaves.

12. *The inverse lattice*

Suppose that one does not consider the leaves as geometrical points but as described by some function on the surface of a cylinder, or in the plane obtained by unrolling the surface of the cylinder. The points of the cylinder may be expressed in the form

$$(\xi_1, \xi_2) \begin{pmatrix} a & b \\ c & d \end{pmatrix} = (x, z)$$

since the matrix

$$\begin{pmatrix} a & b \\ c & d \end{pmatrix}$$

is non-singular. When the function is described in terms of the variables (x, z) it is periodic with unit period in both variables, i.e. of the form

$$\sum A_{m,n} e^{2\pi i(m\xi_1 + n\xi_2)}.$$

To express this in terms of the original variables one must express (ξ_1, ξ_2) in terms of (x, z). If the inverse of $\begin{pmatrix} a & b \\ c & d \end{pmatrix}$ is $\begin{pmatrix} A & B \\ C & D \end{pmatrix}$ then

$$\xi_1 = Ax + Bz, \qquad \xi_2 = Cx + Dz, \tag{I.12.1}$$

and the function can be written as

$$\sum A_{m,n} e^{2\pi i((mA + nC)x + (mB + nD)z)}$$

or as

$$f(x) = \sum Au\, e^{2\pi i(u, x)} \tag{I.12.2}$$

where the summation is over the lattice described by the matrix $\begin{pmatrix} A & B \\ C & D \end{pmatrix}$. This lattice may be called the inverse lattice, because it is described by the transposed inverse of the matrix describing the lattice arising from the congruences.

In the inverse lattice it is only the lattice points relatively close to the origin that are of any particular importance. Consider for example a function in the plane having the symmetry of the lattice, and of the form $\sum_{y \in L} g(x + y)$ where L is the lattice $\begin{pmatrix} a & b \\ c & d \end{pmatrix}$. If

$$g(x) = \int \phi(u)\, e^{2\pi i(u, x)}\, du$$

then the coefficients Au are proportional to $\phi(u)$.

If mJ and nJ are two parastichy numbers, if the corresponding vectors generate the lattice, and if the parastichies make angles ψ_1, ψ_2 with the generators of the cylinder, then the matrix

$$\frac{J}{2\pi\varrho}\begin{pmatrix} n & n\tau_2 \\ -m & -m\tau_2 \end{pmatrix}$$

is one of the matrix descriptions of the inverse lattice. Here

$$\tau_1 = \tan \psi_1, \qquad \tau_2 = \tan \psi_2. \tag{I.12.3}$$

It will be seen that the coefficients A, C are bound to be multiples of $J/2\pi\varrho$. In other words, the first coordinates of the points of the inverse lattice are all multiples of $J/2\pi\varrho$. This simply represents the fact that the leaf pattern is unaltered by rotating the cylinder through the angle $2\pi/J$.

When drawing diagrams to describe a phyllotaxis through its inverse lattice, it is helpful to draw a number of vertical lines, $u = m/2\pi\varrho$ or possibly only the lines $u = mJ/2\pi\varrho$. The points of the inverse lattice are bound to lie on these lines, and may be imagined as beads sliding up and down on them. Owing to the fact that only relatively few points of the inverse lattice can be of importance it is appropriate to use a relatively large scale for such diagrams, i.e. a larger area of paper may be used per lattice point in the case of an inverse lattice than would be appropriate for the primary lattice. It may be noticed that, apart from scale, the two lattices are obtainable from one another by rotating through 90°; the scale factor is Δ or Δ^{-1}.

In part II ideas involved in the inverse lattice will be found of immense importance. To a large extent, however, it will no longer be possible to work in terms of lattices. To assume that one has a lattice is an approximation which is no longer appropriate when discussing the origin of the phyllotactic patterns. It is nevertheless still appropriate to describe functions on the surface of the cylinder by a Fourier analysis of some kind. The appropriate kind of Fourier analysis for functions defined on a cylinder is of the form of a Fourier series in one variable and a Fourier integral in the other:

$$f(x, y) = \sum_{m=-\infty}^{\infty} \int_{-\infty}^{\infty} F_m(\upsilon)\, e^{(imx/P) + i\upsilon y}\, dy. \tag{I.12.4}$$

The function $F_m(\upsilon)$ may be regarded as being defined on the lines $u = m/2\pi\varrho$ of the diagram mentioned above. In general $F_m(\upsilon)$ will be complex, and cannot be very easily represented on the diagram. It may happen however (and in practice it always happens), that $f(x, y)$ has a centre of symmetry, $f(x, y) = f(x_0 - x, y_0 - y)$. In this case, if (x_0, y_0) is made the origin, the function $F_m(\upsilon)$ is real. If in addition, as often happens, the function is nowhere negative, the function may be conveniently shown diagrammatically by representing the vertical lines of varying thickness. The case of a lattice arises when this widening of the lines is restricted to isolated points.

The diagrams for the inverse lattice also include a circle whose radius is $K_0 = 2\pi/\lambda_0$ where λ_0 is the so-called "optimum wavelength". K_0 may be called the "optimum radian wave number". There is a tendency for the lattice points with the greatest coefficients in (I.12.2) to lie not very far from this circle.

The expression (I.12.2) is a very familiar one when applied to three-dimensional lattices. It then gives the relation between the electron density in a crystal and the X-ray reflection pattern. The similarity of phyllotactic patterns to crystal lattices was first observed by the brothers Bravais.

13. *Flow matrices*

If a lattice is changing, the manner of its change may be described by some matrix description $A(t_0)$ of the lattice at some time t_0, together with the product $(A(t))^{-1}A'(t)$ at other times, the dash here representing differentiation. This product will be called the "flow matrix". It is independent of the matrix description chosen, for if $B(t)$ is another matrix description there is an improper unimodular matrix L such that $B(t) = LA(t)$, and if $A(t)$ and $B(t)$ are continuous, L must be constant. But then

$$(B(t))^{-1}B'(t) = (LA(t))^{-1}LA'(t) = (A(t))^{-1}A'(t).$$

If one uses helical coordinates the matrix description may be written as

$$\begin{pmatrix} 2\pi\varrho J^{-1} & 0 \\ \alpha\varrho & \eta \end{pmatrix}$$

and the flow matrix is

$$F = \begin{pmatrix} F_{11} & F_{12} \\ F_{21} & F_{22} \end{pmatrix} = \begin{bmatrix} \dfrac{d\log\varrho}{dt} & 0 \\ \dfrac{\varrho}{\eta}\dfrac{d\alpha}{dt} & \dfrac{d\log\eta}{dt} \end{bmatrix}. \tag{I.13.1}$$

A convenient way of picturing flow matrices is to imagine the change in the lattice as being due to the leaves being carried over the surface of the lattice by a fluid whose velocity is a linear function of position. The flow matrix then gives the relation between the velocity and the position. This point of view is particularly suitable when one is concerned with leaves which are sufficiently mature to be no longer moving with respect to the surrounding tissue, but only have movement due to the growth of that tissue. The coefficient F_{11} then represents the exponential rate of growth in girth of the stem, and the coefficient F_{22} the exponential rate of increase of the stem in length. The sum of these, the trace of the flow matrix, is the exponential rate of increase of the leaf area. The coefficient F_{21} represents whatever tendency there is for the stem to twist. It should be small, or in other words the divergence angle should not be appreciably affected by such growth. If this coefficient F_{21} is zero, the flow may be described as being "without twist". A flow without twist and with $F_{11} = F_{22}$, i.e. a scalar flow matrix, may be described as a "compression". One with $F_{11} + F_{22} = 0$ may be described as "area preserving". One may also consider flows with $F_{12} \neq 0$ but these of course cannot apply to phyllotactic lattices but only to more general lattices. A flow with $F_{11} = F_{22} = 0$, $F_{12} + F_{21} = 0$ represents a rotation.

If a lattice has a representation which changes continuously with time from $A(t_1)$ to $A(t_2)$, then the ratio $(A(t_1))^{-1}A(t_2)$ may be called the "finite flow matrix" for the period t_1 to t_2. Finite flows are also independent of the representation.

If a flow matrix is independent of time, the corresponding finite flow may be expressed as the exponential of the product of the flow and the time for which it acts. Particular cases of exponentials of matrices that may be relevant in this connection are*

$$\exp\begin{pmatrix} 0 & \theta \\ -\theta & 0 \end{pmatrix} = \begin{pmatrix} \cos\theta & \sin\theta \\ -\sin\theta & \cos\theta \end{pmatrix}, \tag{I.13.2}$$

$$\exp\begin{pmatrix} \kappa_1 & 0 \\ 0 & \kappa_2 \end{pmatrix} = \begin{pmatrix} e^{\kappa_1} & 0 \\ 0 & e^{\kappa_2} \end{pmatrix}, \tag{I.13.3}$$

$$\exp\begin{pmatrix} 0 & 0 \\ \chi & 0 \end{pmatrix} = \begin{pmatrix} 1 & 0 \\ \chi & 1 \end{pmatrix}. \tag{I.13.4}$$

They give the effects of constant rotations, twistless flows and pure twists.

If a finite flow matrix G arises from the continuous change of a *phyllotactic* lattice then $G_{12} = 0$ since the point $(2\pi\varrho, 0)$ must transform into another of the same form. Also G_{11} and G_{22} must be positive, since they are continuous, initially both unity, and can never vanish without the

---

* If A is a matrix, $\exp(A) = I + A + A^2/2 + \cdots + A^n/n! + \cdots$. The formula (I.13.2) follows from

$$\begin{pmatrix} 0 & \theta \\ -\theta & 0 \end{pmatrix}^n = \begin{cases} \begin{pmatrix} 0 & \theta^n \\ -\theta^n & 0 \end{pmatrix}(-1)^{(n-1)/2} & \text{if } n \text{ is odd,} \\ \begin{pmatrix} \theta^n & 0 \\ 0 & \theta^n \end{pmatrix}(-1)^{n/2} & \text{if } n \text{ is even.} \end{cases}$$

leaf area vanishing. On the other hand, every matrix G with $G_{12}=0$, $G_{11}>0$, $G_{22}>0$ is a possible finite flow matrix, as may be seen by writing it in the form

$$\begin{pmatrix} G_{11} & 0 \\ 0 & G_{22} \end{pmatrix} \begin{pmatrix} 1 & 0 \\ G_{21}/G_{11} & 1 \end{pmatrix}.$$

It can now easily be seen that if a continuous change of a phyllotactic system is described by a finite flow matrix equal to the matrix G then the corresponding curve in the space of lattices is deformable to zero; for the set of points satisfying $G_{12}=0$, $G_{11}>0$, $G_{22}>0$ is simply connected.

14. *The touching circles phyllotaxis*

The lattice patterns which arise from packing circles as tightly as possible on the surface of a cylinder have been considered as models for phyllotaxis (v. ITERSON 1907). The cylinder is supposed to increase in diameter and the lattice continuously adjusts itself, without major alteration, to allow as many circles as possible per unit area. It is not difficult to see that in tightest packing every circle is touching two others. If a formal proof is desired, one may argue as follows. If no vectors of the lattice are as short as the diameter l of the circles, then η may be decreased until one of them has this length. Thereafter one may continue to decrease η until a second vector has length l, but during this second phase of the process the divergence angle must be continually modified to ensure that the first vector remains of length l. Such a lattice can be described by the matrix

$$l\begin{pmatrix} \sin\psi_1 & \cos\psi_1 \\ \sin\psi_2 & \cos\psi_2 \end{pmatrix}$$

where

$$|\psi_1|<\pi/2, \qquad |\psi_2|<\pi/2, \qquad \pi/3 \leqslant \psi_1-\psi_2 < 2\pi/3$$

and the flow matrix will be found to be

$$\frac{d\log\varrho}{dt}\begin{pmatrix} 1 & 0 \\ -(\tan\psi_1+\tan\psi_2) & \tan\psi_1\tan\psi_2 \end{pmatrix}.$$

The angles ψ_1, ψ_2 are obliged to satisfy

$$\dot\psi_1\tan\psi_1 = \dot\psi_2\tan\psi_2$$

in order that $F_{12}=0$.

At certain values of the radius the third principal vector is also of length l. The lattice may then be called "equilateral". For such a lattice $\psi_1-\psi_2$ must be either $\pi/3$ or $2\pi/3$. It may be supposed to be $2\pi/3$; otherwise one may interchange the first two vectors and change the sign of one of them. If one then puts $\psi_3 = \psi_2 + 2\pi/3$, all the angles $\psi_1 - \psi_2$, $\psi_2 - \psi_3$, $\psi_3 - \psi_1$ will be $2\pi/3$. If the radius of the cylinder is allowed to alter still further, the directions of the three sides of the principal triangle will still be given by the angles ψ_1, ψ_2, ψ_3 or angles very close to them, but their roles will have been interchanged. The first two vectors will be in directions ψ_1' and ψ_2' and the third in direction ψ_3' and

$$\psi_1' - \psi_2' = \psi_2' - \psi_3' = \psi_3' - \psi_1' = 2\pi/3$$

and $\psi_1', \psi_2', \psi_3'$ will be a permutation of ψ_1, ψ_2, ψ_3. The permutation has to be cyclic to ensure $\psi_1' - \psi_2' = \psi_2' - \psi_3' = \psi_3' - \psi_1' = 2\pi/3$. The question as to which of the three angles ψ_1, ψ_2, ψ_3 is to be ψ_3' must be decided by the condition that the subsequent change of the lattice shall, for a short time at any rate, satisfy $\pi/3 \leqslant \psi_1' - \psi_2' < 2\pi/3$ and, subject to this, the area $\Delta = l^2 \sin(\psi_1' - \psi_2')$ shall be as small as possible. This means to say that $d\Delta/dt$ must change sign but, subject to this, have the smallest available modulus. Now in the period before the change

$$\frac{1}{\Delta} \frac{d\Delta}{dt} = (1 + \tan \psi_1 \tan \psi_2) \frac{d \log \varrho}{dt}$$

$$= \frac{\cos(\psi_1 - \psi_2)}{\cos \psi_1 \cos \psi_2 \cos \psi_3} \cos \psi_3 \frac{d \log \varrho}{dt}$$

$$= \frac{-d \log \varrho/dt}{2\cos \psi_1 \cos \psi_2 \cos \psi_3} \cos \psi_3$$

whereas afterwards

$$\frac{1}{\Delta} \frac{d\Delta}{dt} = \frac{-d \log \varrho/dt}{2\cos \psi_1' \cos \psi_2' \cos \psi_3'} \cos \psi_3'.$$

Then $\cos \psi_3'$ has the opposite sign to $\cos \psi_3$ and $|\cos \psi_3'|$ must equal the smaller of $|\cos \psi_1|$ and $|\cos \psi_2|$. Since $\cos \psi_1 + \cos \psi_2 + \cos \psi_3 = 0$ the change of sign is certainly possible. In the case that $\cos \psi_1$ and $\cos \psi_2$ have the same sign,

$$|\cos \psi_3'| = \text{Min} |\cos \psi_r|$$

whereas if they have the opposite sign

$$|\cos \psi_3'| = \text{Max} |\cos \psi_r|.$$

In no case is $|\cos \psi_3'|$ intermediate between $|\cos \psi_1'|$ and $|\cos \psi_2'|$. But the three principal parastichy numbers (after the change) are proportional to $|\cos \psi_1'|$, $|\cos \psi_2'|$, $|\cos \psi_3'|$. The third principal parastichy number is therefore either the smallest or the greatest.

⟦78⟧

In the case that $d \log \varrho/dt > 0$, $\cos \psi_1' \cos \psi_2' < 0$ (since $d\Delta/dt > 0$) and the third parastichy number is the smallest. Then the hypothesis of geometrical phyllotaxis is satisfied immediately after the lattice is equiangular. It will continue to be so until there is a change in the principal parastichy numbers. The first two can only change at equiangular lattices but the third can also change when the lattice is square. Now if q_1, q_2 are the first two parastichy numbers the third is either $q_1 + q_2$ or $|q_1 - q_2|$. If the change is upward, i.e. if $d \log \varrho/dt > 0$, and the third parastichy number is the smallest, the new value of the third parastichy number must be $q_1 + q_2$ and the hypothesis is still satisfied. If the parastichy number is decreasing, the new value of the third parastichy number is $|q_1 - q_2|$ and the hypothesis will be satisfied if and only if q_1/q_2 lies between $1/2$ and 2. Thus the hypothesis of geometrical phyllotaxis is satisfied in the case of "touching circles phyllotaxis" from the point when the lattice first becomes equiangular and can only cease to do so when $d\varrho/dt < 0$, and the third parastichy number cannot be decreased to become the difference of the first two without contradicting the hypothesis, therefore

In a continuously varying touching circles phyllotaxis the hypothesis of geometrical phyllotaxis is satisfied from the time when an equilateral triangle first appears onwards.

The difficulty in the proof given above lay largely in deciding which side of the equilateral triangle increases when the diameter of the cylinder increases. The following not very rigorous argument may be found helpful. Consider three of the circles forming an equilateral triangle of the lattice (Fig. 4). The circles are being pressed downwards to ensure the closest packing. The downward pressure of the upper circle will tend to wedge the lower circles apart, whilst at the same time holding it in contact with the other two. Thus if ϱ is increasing it is the most nearly horizontal of the sides which increases. When a touching circles lattice with decreasing ϱ reaches a state when $\tan \psi_1 \tan \psi_2 = \infty$ it is no longer possible for the lattice to continue. This happens if ψ_1 or ψ_2 is 90°.

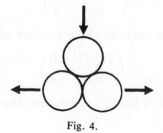

Fig. 4.

It is of some interest to know what the angles ψ_1, ψ_2, ψ_3 will be when the lattice is equilateral. Writing $l \cos \psi_i = \eta m_i$ one obtains

$$\tan \psi_i = (m_{i+1} - m_{i-1})/m_i \sqrt{3}$$

(the suffixes being reckoned mod 3). When the parastichy numbers are 1, 1, 2 the directions are $0°$ mod $60°$. When they are 1, 2, 3 the directions are $10°54'$ mod $60°$. When the parastichy numbers are 2, 3, 3 the directions are $6°35'$ mod $60°$. For the limit of large parastichy numbers of the Fibonacci series the directions are $\tan^{-1}(\omega^{-3}/\sqrt{3})$, i.e. $7°46'$ mod $60°$. It will be seen that these lattices have vectors lying very nearly along the generators of the cylinder. This will be of importance in §18, where evidence will be brought to show that the touching circles theory is unlikely to be valid. The arguments apply also equally well to discredit any theory requiring that for some values of the radius of the cylinder the lattice be equiangular.

The divergence angles for the equiangular lattices are as follows:

1,2,2	$180°$
1,2,3	$128°34'$
2,3,5	$142°6'$
F_{n-1}, F_n, F_{n+1}	$2\pi \left(\dfrac{F_{n-1}F_{n+1} + F_{n-1}^2 + F_{n-2}^2}{F_{n+1}^2 + F_n^2 + F_{n-1}^2} \right).$

Closely related to touching circles phyllotaxis is a phyllotaxis whose inverse lattice is a touching circles lattice. The importance of such a lattice is that its points may be regarded as the maxima of two sets of waves of fixed wavelength superimposed on one another.

15. *The lattice described by its twist and other coordinates*

Any phyllotactic lattice may be described by its helical coordinates in the form of the matrix

$$\begin{pmatrix} 2\pi\varrho J^{-1} & 0 \\ \alpha\varrho & \eta \end{pmatrix}.$$

Alternatively, it may be expressed as a product of the matrix

$$\begin{pmatrix} 2\pi\varrho J^{-1} & 0 \\ \alpha_0\varrho & \eta \end{pmatrix}$$

describing a lattice with a limiting divergence angle α_0, and a finite flow matrix

[80]

$$\begin{pmatrix} 1 & 0 \\ \chi & 1 \end{pmatrix}$$

describing a twist. The matrix describing the lattice may be further broken down into factors as follows

$$\Delta^{1/2} \begin{pmatrix} 1 & 0 \\ \alpha_0 J/2\pi & 1 \end{pmatrix} \begin{pmatrix} \kappa & 0 \\ 0 & \kappa^{-1} \end{pmatrix} \begin{pmatrix} 1 & 0 \\ \chi & 1 \end{pmatrix} \qquad \text{(I.15.1)}$$

where $\kappa = (2\pi\varrho/\eta J)^{1/2} = 2\pi\varrho/J\Delta^{1/2}$. One may of course multiply it on the left by any unimodular matrix, for instance

$$\begin{pmatrix} p_n & q_n \\ p_{n+1} & q_{n+1} \end{pmatrix}$$

where p_n/q_n and p_{n+1}/q_{n+1} are two successive convergents of $\alpha_0 J/2\pi$.

Since $\alpha_0 J/2\pi$ is a limiting divergence angle, the partial quotients of $\alpha_0 J/2\pi$ are all 1 beyond some point and it will be supposed that p_n/q_n and p_{n+1}/q_{n+1} are obtained by taking sufficiently many partial quotients to ensure that some of these 1's are included. Then

$$q_n \frac{\alpha J}{2\pi} - p_n = (-1)^n \omega^{-n} A \quad \text{where} \quad A = \frac{\omega J}{q_0 + \omega q_1}$$

and the lattice is described by the matrix

$$\Delta^{1/2} \begin{pmatrix} A\kappa(-\omega^{-1})^n & \kappa^{-1} q_n \\ A\kappa(-\omega^{-1})^{n+1} & \kappa^{-1} q_{n+1} \end{pmatrix} \begin{pmatrix} 1 & 0 \\ \chi & 1 \end{pmatrix}. \qquad \text{(I.15.2)}$$

In the case that $\chi = 0$, for some n, Jq_n, Jq_{n+1}, Jq_{n-1} are the principal parastichy numbers, and (I.15.2) is the principal matrix representation. In this way the lattice is described by the leaf area Δ, the parameter κ which is directly related to the plastochrone ratio, and the twist χ. The parastichy series is supposed known so that the value of $A = \omega J/(q_0 + q_1\omega)$ is determined. In theory one can refer the phyllotaxis to any parastichy series, but in practice if one refers it to the "wrong" series monstrously large twists are required.

For large parastichy numbers, q_n is approximately $(q_0 + \omega q_1)\omega^{n-1}/\sqrt{5}$ i.e. $J\omega^n/A\sqrt{5}$, and the matrix approximates the form

$$\frac{\Delta^{1/2}}{5^{1/4}} \begin{pmatrix} \omega^{2\theta+1/2} & \omega^{-2\theta-1/2} \\ -\omega^{2\theta-1/2} & \omega^{-2\theta+1/2} \end{pmatrix} \begin{pmatrix} 1 & 0 \\ \chi & 1 \end{pmatrix} \qquad \text{(I.15.3)}$$

where n has been supposed even and $\theta = -n/2 - 1/4 + \frac{1}{2}\log_\omega(A\kappa)$. Lattices of the form (I.15.3) with $\chi = 0$ may be called "ideal phyllotactic lattices". Apart from the leaf area, which is only a scale factor, the only parameter is θ, and the lattice is a periodic function of θ with unit period. If $1/2$ is added to θ the lattice is transformed into the mirror image in a generator of the cylinder for

$$\begin{pmatrix} \omega^{2\theta+3/2} & \omega^{-2\theta-3/2} \\ -\omega^{2\theta+1/2} & \omega^{-2\theta-1/2} \end{pmatrix} = \begin{pmatrix} -1 & 1 \\ 1 & 0 \end{pmatrix} \begin{pmatrix} \omega^{2\theta+1/2} & \omega^{-2\theta-1/2} \\ -\omega^{2\theta-1/2} & \omega^{-2\theta+1/2} \end{pmatrix}.$$

A very natural way of describing a general lattice is to express it by means of a matrix which is the product of one which describes an ideal lattice and another matrix of unit determinant and with two parameters, thus for instance

$$\frac{\Delta^{1/2}}{5^{1/4}} \begin{pmatrix} \omega^{2\theta+1/2} & \omega^{-2\theta-1/2} \\ -\omega^{2\theta-1/2} & \omega^{-2\theta+1/2} \end{pmatrix} \begin{pmatrix} 1 & \mu \\ 0 & 1 \end{pmatrix} \begin{pmatrix} 1 & 0 \\ \chi & 1 \end{pmatrix}. \qquad (I.15.4)$$

The order of the matrices $\begin{pmatrix} 1 & 0 \\ \chi & 1 \end{pmatrix}$ and $\begin{pmatrix} 1 & \mu \\ 0 & 1 \end{pmatrix}$ is conveniently immaterial. As a finite flow, the first of these represents a pure twist and the second a pure shear. The shear coordinate is closely related to the plastochrone ratio by the relation

$$\mu = \pm(-1)^n \begin{vmatrix} q_{n-1} & q_n \\ q_n & q_{n+1} \end{vmatrix} \frac{\Delta^2}{2\pi\varrho}$$

which holds for any n, with sign independent of n.

A convenient way of describing the effect of a finite twisting flow

$$\begin{pmatrix} 1 & 0 \\ \chi & 1 \end{pmatrix}$$

is to say that it adds χ to the tangent of all the angles which lattice vectors make with the generators of the cylinder. This idea at once provides a method of determining the twist of a phyllotactic system. Suppose that the parastichies with parastichy numbers q_n, q_{n+1} make angles ψ_n, ψ_{n+1} with the generators, and that the limiting angle concerned is that which corresponds to the sequence with q_n, q_{n+1} as consecutive terms. Then if the relation $\tan \psi_n = -\omega \tan \psi_{n+1}$ were satisfied, the divergence angle would have the limiting angle. Since this relation can be ensured by subtracting χ from both $\tan \psi_n$ and $\tan \psi_{n+1}$, the equation

$$\tan \psi_n - \chi = -\omega(\tan \psi_{n+1} - \chi)$$

must hold, i.e.

$$\chi = (\tan \psi_n + \omega \tan \psi_{n+1})/(1+\omega).$$

In the case of an equilateral lattice of high parastichy number, one may take $\tan \psi_n = (3\omega^{-1})/2\sqrt{3}$, $\tan \psi_{n+1} = \omega^{-3}/\sqrt{3}$, $\chi = 5/(2\omega^3\sqrt{3}) = 0.341$.

In the case of an ideal lattice with $\theta = 0$, the matrix is orthogonal and so represents a square lattice, which is then also a touching circles lattice. Thus the touching circles lattice is ideal when both are square. The sides of the square make an angle $\tan^{-1}\omega^{-1}$, i.e. 31°43′ with the coordinate axes.

16. *The optimum packing problem*

If one submits a lattice, with matrix description

$$\begin{pmatrix} a & b \\ c & d \end{pmatrix}$$

to an area preserving twistless flow

$$\begin{pmatrix} \kappa & 0 \\ 0 & \kappa^{-1} \end{pmatrix}$$

then the points of the lattice all move along rectangular hyperbolae, and have a non-zero minimum distance from the origin. The point

$$(m, n)\begin{pmatrix} a & b \\ c & d \end{pmatrix}$$

will have minimum distance $[2(ma+nc)(mb+nd)]^{1/2}$. The minimum of this distance taken over all the points of the lattice is of a certain interest and importance. It is the shortest distance to which any leaves approach one another during the flow. If this distance is small in comparison with $\Delta^{1/2}$ then the leaves will, at some stage of the flow, become awkwardly close. It is interesting in this connection that *the ideal lattices are optimum in the sense that with them the minimum distance has the maximum value for the given value of Δ.* To prove this, let

$$\begin{pmatrix} a & b \\ c & d \end{pmatrix}$$

be a matrix representation of the lattice in which $ac<0$, $bd>0$. Such a representation can be made from two of the principal vectors. Then the minimum of $|(ma+nc)(mb+nd)|$ for pairs of integers m, n of the same sign (say for positive m, n) occurs when m/n is one of the convergents of the continued fraction of $-c/a$. If not, let $p_r/q_r, p_{r+1}/q_{r+1}$ be two consecutive convergents such that $q_r \leqslant n \leqslant q_{r+1}$. Since the matrix

$$\begin{pmatrix} p_r & q_r \\ p_{r+1} & q_{r+1} \end{pmatrix}$$

has determinant ± 1, one can find integers k, l such that $m = kp_r + lp_{r+1}$,

$n = kq_r + lq_{r+1}$. Since m/n is not a convergent, neither k nor l can be zero. They must be of opposite sign since $n < q_{r+1}$. Now if $m \geqslant p_r$ then

$$|mb + nd| > |p_r b + q_r d|$$

and

$$|ma + nc| = |k(p_r a + q_r c) + l(p_{r+1} a + q_{r+1} c)| > |p_r a + q_r c|.$$

But then m, n could not then give the minimum for p_r, q_r would do better. Then $m < p_r$ and $n > q_r$. But in this case by reducing q_r by 1 one would certainly reduce $|mb + nd|$ and one would reduce $|ma + nc|$ for

$$ma/c + n > ma/c + (n-1) \geqslant (-a/c)(p_r + q_r a/c + 1) \geqslant 0.$$

This shows that m/n must be a convergent of $-c/a$. Likewise if m/n is negative it must be a convergent of $-d/b$. In the case of the ideal lattice, $-a/c = \omega$, $b/d = \omega^{-1}$ and m and n are consequently Fibonacci numbers. Then

$$|F_r a + F_{r+1} c| = |c| \, |\omega F_r - F_{r+1}| = |c| \, \omega^{-r},$$

$$|F_r b + F_{r+1} d| = |b| \, |\omega F_r + F_{r+1}| = b\omega^{r+1} = |d| \, \omega^r,$$

$$|F_r a + F_{r+1} c| \, |F_r b + F_{r+1} d| = |cd| = \Delta/\sqrt{5}.$$

Thus the shortest distance is

$$5^{-1/4}(2\Delta)^{1/2} = 0.8945\sqrt{\Delta}.$$

When a lattice of such a family has worst packing then $\pm(ma + nc) = mb + nd$ and $(ma + nc, mb + nd)$ is a principal vector, i.e. one may take $a = b$ and $m = 1$, $n = 0$. The worst packing ideal lattice is

$$\left(\frac{\Delta}{\sqrt{5}}\right)\begin{pmatrix} 1 & 1 \\ -\omega & \omega^{-1} \end{pmatrix}.$$

It must still be shown that no other lattice gives so large a minimum. Write $A_r = p_r a + q_r c$, $B_r = p_r b + q_r d$, then the determinants $A_r B_{r-1} - A_{r-1} B_r$ all have the value $\pm\Delta$. If the partial quotients of $-a/c$ are K_r, then

$$A_{r+1} = K_{r+1} A_r + A_{r-1}, \qquad B_{r+1} = K_{r+1} B_r + B_{r-1}.$$

Now suppose $K_{r+1} \geqslant 3$, then for $B_r/B_{r+1} > 0$, $A_r/A_{r+1} < 0$

$$\frac{|\Delta|}{|A_r B_r|} = \left| \frac{B_{r+1}}{B_r} - \frac{A_{r+1}}{A_r} \right| \geqslant 3.$$

Since

$$\frac{|\Delta|}{|A_r B_r|} < \sqrt{5} < 3$$

it follows that there can be no partial quotients as large as 3 in the partial fraction for $-c/a$. Then $B_{r+1}/B_r \leqslant 3$ for each r, and so if $K_r = 2$ then

$$\frac{|\Delta|}{|A_r B_r|} \geqslant \left| \frac{B_r}{B_{r+1}} \right| = \left| K_r + \frac{B_{r-1}}{B_r} \right| > 2\tfrac{1}{3} > \sqrt{5}.$$

There can therefore not be any partial quotients as large as 2, so they are all 1, i.e. $-c/a$ is ω or ω^{-1}. Likewise b/d is ω or ω^{-1}. [There is a handwritten marginal note here that "more argument is needed for $r = 0$". The above proof is indeed not valid for K_0, but the value of K_0 does not affect the divergence angle.]

Though this optimum property is of considerable mathematical interest, its biological importance is perhaps rather secondary. Above all it would be quite unjustified to suppose that the appearance in nature of nearly ideal lattices is due to a search for the best lattice. On the contrary it seems probable that the effect of such a search would be to defeat its own ends. It would be likely that the evolutionary process would lead to some not too bad lattice which was a local optimum and remain there. It will be realised that if the ratio a/c is allowed to change continuously it has to pass through rational values, and these are the very worst from the point of view of packing. More specifically, if one wishes to alter one of the partial quotients of a number by altering the number continuously, then it is necessary to allow the next partial quotient to take unlimited large values. Of course not all the partial quotients concerned in a phyllotactic lattice can be of importance, but if any one of them is of sufficient importance that it must be kept down to a moderate size then this fact prevents any of the previous partial quotients from being altered. However, although it is unreasonable to suppose that there is any such evolutionary search for the best lattice from the point of view of packing in spite of twistless area-preserving flow, the fact that the naturally occurring lattices have, or very nearly have, the optimum property, still has its advantages. It means in effect that if there are mutations which modify the twistless growth, disadvantageous packing effects will arise.

17. *Comparison of methods of describing lattices*

A considerable number of different sets of parameters have now been introduced for the description of phyllotactic lattices. Their various purposes, merits and defects will now be compared.

(1) The most fundamental way of describing lattices is by a matrix

$$\begin{pmatrix} a & b \\ c & d \end{pmatrix}.$$

The other methods described may all be related to it. Its main advantage is its generality, and its main disadvantage its lack of uniqueness.

(2) Closely related to the matrix describing the lattice is that which describes the inverse lattice

$$\begin{pmatrix} A & B \\ C & D \end{pmatrix}.$$

This will be found particularly useful in part II.

(3) Amongst methods of making the matrix description unique is the use of helical coordinates, i.e. the use of the matrix description

$$\begin{pmatrix} 2\pi\varrho/J & 0 \\ \alpha\varrho & \eta \end{pmatrix}.$$

If the plastochrone distance η is required to be positive and the modulus of the divergence angle α, satisfies $-\pi < \alpha \leqslant \pi$, this representation is unique. This form of description is more suitable for the description of the lattice as a group of congruences of the cylinder, but is not very helpful where theories of the origin of the phyllotactic pattern are concerned. The divergence angle and the plastochrone displacement are not easily measured or even appreciated on specimens with high parastichy numbers. As compared to the method next to be described the helical coordinates at least have the advantage of changing continuously with continuously changing lattices.

(4) Another method of making the matrix description unique is to make the vectors represented by the two rows of the matrix be the first two principal vectors. This is the principal matrix representation. This representation is, one might say, the most natural matrix representation, i.e. that which one would be most likely to choose if asked to give a matrix representation of a lattice. Its main disadvantage is that it undergoes discontinuous changes when the lattice changes continuously.

(5) There are methods of describing a lattice by means of parameters which vary continuously with change of lattice, and in such a way that lattices whose principal matrix representations are near to one another are represented by neighbouring sets of parameters. Such descriptions can for instance be based on the theory of elliptic functions. The disadvantage of these methods is that these sets of parameters are most unmanageable from the point of view of their algebraic properties. They are not further discussed elsewhere in this paper.

(6) When one wishes to measure the parameters of a lattice, suitable quantities are the radius ϱ and the two angles ψ_1, ψ_2 which the two

parastichies make with the generators. In addition to these measured quantities one needs to know the relevant parastichy numbers.

(7) When one is concerned with phyllotactic lattices belonging to a known series of parastichy numbers, rather different parameters are appropriate. These are the parameters $\Delta, \theta, \mu, \chi$ (expression I.15.4). These parameters vary continuously in a continuously varying lattice. In theory it is possible for very similar lattices to be described by very different parameters, but this does not cause any genuine misunderstanding.

18. *Variation principle theories. Equilateral lattices*

One rather attractive type of theory to account for the change of phyllotaxis with changing radius is to suppose that there is some function of the lattice that the plant attempts to minimise. It is only able to achieve local minima, and is restricted to phyllotactic lattices that can be fitted to the cylinder available. This "potential" function should of course be defined for lattices other than phyllotactic lattices, and should be unchanged on rotating the lattice. The touching circles phyllotaxis can be defined by such a potential function, viz. the ratio of the length of the shortest lattice vector to l, so long as this ratio exceeds 1, and by $\Delta/2l$ otherwise. Likewise the fixed wavelength lattices can be defined by a similar potential. Other potential functions may be defined in the form

$$\phi(\Lambda) = \sum_{u \in \Lambda} f(|u|).$$

The function $f(r)$ should preferably tend to zero quickly as r tends to infinity, and have a negative minimum at some positive value, l. When the minimum is very sharp, one approaches the touching circles lattices again.

Nearly all such theories require that at certain values of the plastochrone ratio the lattice must be hexagonal, for if the equilateral lattice is optimum with a certain vector length in the infinite plane it will also be optimum on any circle on which it can be fitted. Now suppose that the vector length is chosen so that a lattice (in the unrestricted plane) is equilateral and gives the minimum potential for equilateral lattices. Then although this lattice may not be an optimum or even a local optimum it is at least locally stationary in the space of lattices. For suppose the lattice begins to change with flow matrix F. Then the rate of change of potential will be linear in the coefficients of F, e.g. trace(FH). Now if the whole lattice is rotated by a matrix U then F becomes UFU^{-1}. Since for rotations of 60° the rate of change of potential due to the flow F will be unaltered by the rotation, it follows that, if U is such a rotation, trace$(UFU^{-1}H)=$ trace(FH) for any

[[87]]

F, i.e. $U^{-1}HU = H$. Thus H commutes with the matrices

$$\begin{pmatrix} \cos r\pi/3 & \sin r\pi/3 \\ -\sin r\pi/3 & \cos r\pi/3 \end{pmatrix}$$

and is therefore of the form

$$\begin{pmatrix} A & B \\ -B & A \end{pmatrix}.$$

But since the potential is stationary for pure compressions, $A = 0$. It is also stationary for pure rotations, and therefore $B = 0$. Hence the equilateral lattice is a stationary point of the potential. It may not be a minimum, and if it is a minimum it may still not be one which really ever comes into play. But these possibilities on the whole seem rather unlikely. One must expect that theories depending on a variation principle will involve equilateral lattices for appropriate radii.

It seems however that no such theory can be right, for in the experimental material there seems to be no trace of any equilateral lattices. As has been mentioned, equilateral lattices of high parastichy number have a twist χ of about 0.34. In actual mature specimens one seldom finds values of χ even as large as 0.1. This might possibly be explained by the lattice being subjected to a squashing flow

$$\begin{pmatrix} \kappa & 0 \\ 0 & \kappa^{-1} \end{pmatrix}$$

in the growth after the lattice has been formed, which results in the twist getting magnified by the factor κ^2. It is difficult however to estimate the values of κ which might apply. But the same applies with growing points.

An equilateral lattice, if it occurred, would be rather unsatisfactory from the point of view of packing. A not very great degree of squashing applied to an equilateral lattice gives a lattice with very poor packing indeed.

Part II. Chemical Theory of Morphogenesis

1. *Morphogen equations for an assembly of cells. The linear case*

In TURING (1952) the theory of a reaction and diffusion system was developed for the case where the geometrical form of the organism was a ring of cells, and where the reaction rates might be considered as linear functions of the concentrations. The equations that will be found in this part are applicable to arbitrary geometrical forms and reaction rate functions. In this investigation, as in the previous one, the geometrical form is assumed to remain unchanged throughout. This assumption cannot of course always be satisfied—indeed variations of chemical concentrations would be of little importance if they did not ultimately affect growth—but the rates of growth are likely to be slow enough for the equilibria of chemical concentration that are reached not to be appreciably affected by the growth.

The description of the organism may be divided into a geometrical and a chemical part, concerned respectively with the diffusion and the reactions of the morphogens which are to be found in it. The word "morphogen", which was introduced in TURING (1952), was there, in effect, defined to mean essentially "chemical substance relevant to morphogenesis". In the present paper it will be given a slightly more restrictive meaning, viz. "chemical substance, the variation of whose concentration is described by a variable in the mathematical theory". The state of the organism at any time t may be described by MN numbers Γ_{mn} ($m = 1, 2, ..., M$; $n = 1, 2, ..., N$), where Γ_{mn} is the concentration of the mth morphogen in the nth cell. This description supposes that there is no need to distinguish one point of a cell from another, an assumption which is probably true, as there is usually considerable protoplasmic flow in the interior of cells, which will result in good mixing of the contents. It is not necessary to assume the cells of equal volume, and one may suppose the rth cell to have volume v_r. The rate of flow from one cell to another will of course be proportional to the difference of concentrations of the flowing substance, i.e. the rate of flow of the mth morphogen from cell r to cell s will be proportional to $\Gamma_{mr} - \Gamma_{ms}$. It must also be proportional to a quantity g_{rs} dependent on the geometry of the wall of separation between cells r and s, but independent of the substance flowing, and there will be a further factor μ_m, the diffusion constant for the morphogen in question, diffusing through the material of which all the cell walls are assumed to be made.

Ignoring the chemical reactions, the equations of the system are

$$v_r \frac{\mathrm{d}\Gamma_{mr}}{\mathrm{d}t} = \mu_m \sum_{s \neq r} g_{rs}(\Gamma_{rs} - \Gamma_{ms}). \tag{II.1.1}$$

If g_{rr} is defined to be

$$- \sum_{s \neq r} g_{rs}$$

then (II.1.1) may be written in the more convenient form

$$v_r \frac{\mathrm{d}\Gamma_{mr}}{\mathrm{d}t} = -\mu_m \sum_s g_{rs}\Gamma_{ms} \tag{II.1.2}$$

or, by putting $\Gamma_{mr}^{(1)} = v_r^{1/2} \Gamma_{mr}$, in the form

$$\frac{d\Gamma_{mr}^{(1)}}{dt} = -\mu_m \sum_s \frac{g_{rs}}{\sqrt{(v_r v_s)}} \Gamma_{ms}^{(1)}. \qquad \text{(II.1.3)}$$

Since the matrix $g_{rs}/\sqrt{(v_r v_s)}$ is symmetrical, it may be brought to diagonal form by an orthogonal transformation:

$$\frac{g_{rs}}{\sqrt{(v_r v_s)}} = \sum_k \alpha_k l_{rk} l_{sk}, \qquad \text{(II.1.4a)}$$

$$\sum_k l_{rk} l_{sk} = \delta_{rs} \qquad \text{(II.1.4b)}$$

and if one then puts

$$\Gamma_{mj}^{(2)} = \sum_r \Gamma_{mr}^{(1)} l_{rj} \qquad \text{(II.1.5)}$$

and consequently

$$\Gamma_{mr}^{(1)} = \sum_j \Gamma_{mj}^{(2)} l_{rj} \qquad \text{(II.1.6)}$$

the equations become simply

$$\frac{d\Gamma_{mj}^{(2)}}{dt} = -\mu_m \alpha_j \Gamma_{mj}^{(2)}. \qquad \text{(II.1.7)}$$

The characteristic values α_k are real, since g_{rs} is real and symmetric. None of them is negative, as can be seen on physical grounds. If one of them were negative, then there would be solutions of (II.1.7) in which concentration differences increase with time. There can be only one α_k which is zero, provided the organism is connected, for there is then, for each m, only one linearly independent solution of the equations which is constant in time.

As regards the chemical reactions, the one essential point is that they proceed at rates which depend only on the concentrations of the various morphogens in the same cell. In TURING (1952) the main interest centred around the case in which the reaction rates are linear functions of the concentrations, an assumption which is reasonably valid so long as only small variations of concentration are concerned. The theory of this linear case was carried through with the rather special geometrical assumption that the organism consisted of a ring of cells. This restriction was, however, an altogether unnecessary one. It was made merely in order to make the problem under consideration a quite definite one, and so make the argument more generally intelligible. As will be seen very shortly, the conclusions which were obtained in that case can be directly taken over to any arrangement of cells.

[90]

Suppose that when the concentrations of the M morphogens are $\Gamma_1, \Gamma_2, \ldots, \Gamma_M$ the rate of production of the mth morphogen is $f_m(\Gamma_1, \ldots, \Gamma_M)$ per unit volume. In this case the equations describing the effect of diffusion and reaction together are

$$v_r \frac{\mathrm{d}\Gamma_{mr}}{\mathrm{d}t} = -\mu_m \sum_s g_{rs}\Gamma_{ms} + v_r f_m(\Gamma_{1r}, \ldots, \Gamma_{Mr}). \qquad \text{(II.1.8)}$$

The equations may be transformed by the substitutions (II.1.3), (II.1.6) to give a result analogous to (II.1.7).

If $\Gamma_{mr}^{(3)}$ are variables which are similar to $\Gamma_{mr}^{(2)}$, except that they refer to differences from the equilibrium, and if these differences are sufficiently small for it to be admissable to treat the reaction rates as linear functions of them, then the transformed equations become

$$\frac{\mathrm{d}\Gamma_{mj}^{(3)}}{\mathrm{d}t} = -\mu_m \alpha_j \Gamma_{mj}^{(3)} + \sum_k a_{mk}\Gamma_{kj}^{(3)} \qquad \text{(II.1.9)}$$

where a_{mk} is the value of $\partial f_m/\partial\Gamma_k$ for the equilibrium concentrations. It will be seen that these equations separate into N independent sets of M equations each. In each set of equations the geometry of the system comes into the problem only through the characteristic values α_j of the diffusion matrix $g_{rs}/\sqrt{(v_r v_s)}$. With the rings of cells considered in TURING (1952) every possible non-negative value of α (there written U) could arise, and was allowed for, and no greater variety of values of α can arise with any other geometrical arrangement.

The solution of (II.1.9) can be written in the form

$$\Gamma_{mj}^{(3)}(t) = \sum_r \Gamma_{rj}^{(4)}(t)S_{mrj} \qquad \text{(II.1.10a)}$$

where

$$\Gamma_{rj}^{(4)}(t) = \Gamma_{rj}^{(4)}(0)\, e^{p_r(\alpha_j)t} \qquad \text{(II.1.10b)}$$

and

$$(p_r(\alpha_j) + \mu_m \alpha_j)S_{mrj} = \sum_s a_{ms}S_{srj}. \qquad \text{(II.1.11)}$$

On eliminating the S_{mrj} from equations (II.1.11) one obtains an algebraic relation between p and α. For each value of α there will be M values of p satisfying the relation, and M corresponding solutions (II.1.10). TURING (1952) was very largely concerned with this relation between α and p, and the various forms it can take with different chemical conditions. It would not be appropriate to repeat that analysis here, but some of the main points of relevance to the present problem may be mentioned. Evidently the terms in (II.1.10b) of greatest importance are those for which the real part of p

is greatest, for these are the ones which grow fastest. Ultimately, in fact, provided that the linearity assumption remains true, one may ignore all components of the solution except those with the largest $\text{Re}(p)$. In TURING (1952) the possibilities were classified according to the values of p and of α when $\text{Re}(p)$ has its maximum value. Since α may be zero, finite or infinite, and p may be real or complex, there are six alternatives, each of which was shown to occur with appropriate imaginary chemical reactions. The case of chief interest, and the only one to be considered here, was where p is real at the maximum, and α is finite and different from zero. This is described as the "case of stationary waves".

When the organism contains only a finite number of cells, or where the cells are infinitesimal but the whole organism of finite volume, the characteristic values α_k will be discrete. There can then only be a finite number for which $\text{Re}(p)$ has its greatest value. It will however be by no means unusual for this number to exceed unity. In fact, if the system has some geometrical symmetry, multiple roots of the characteristic equation of (II.1.11) will not be at all exceptional, for when a symmetry operation is applied to the S_{mrj} they will be converted into characteristic vectors corresponding to the same characteristic value. It may happen that the vector $\Gamma_m^{(3)}$ itself has the symmetry in question, but if it has not (and there is no reason why it should) then the symmetry operation transforms it into *another* solution, and the characteristic root must be multiple. If there is a k-fold root corresponding to the greatest $\text{Re}(p)$, then the limiting solution has k parameters. This situation is greatly modified when the quadratic terms are taken into account.

For many purposes the description of the organism in terms of cells is inconvenient. It may be mathematically more satisfactory to take the limiting case in which the volume of each cell is allowed to shrink to zero and the number of cells is allowed to increase correspondingly. There is no need to assume that the resulting continuous tissue is either homogeneous or isotropic, nor indeed is there any need to carry the theory through in any detail. In any particular case there will be a certain linear operator represented by "∇^2", such that the diffusion of a substance with concentration c and diffusion constant μ obeys the law

$$\frac{dc}{dt} = \mu\nabla^2 c. \qquad (II.1.12)$$

When reactions are taken into account, the equations become

$$\frac{d\Gamma_m}{dt} = \mu_m\nabla^2\Gamma_m + f_m(\Gamma_1, ..., \Gamma_m). \qquad (II.1.13)$$

As in the finite case, one may expand any function of position in the organism in terms of characteristic functions of the operator ∇^2 analogously to (II.1.5), and the remainder of the theory of this section may be applied to this expansion.

2. *Assumptions concerning the chemistry of phyllotaxis*

The behaviour of the solutions (II.1.13) can take very different forms according to the functions f_m and the diffusibilities μ_m involved. This variety is even further increased if the concentrations of the morphogens affect the growth of the organism. It was suggested in TURING (1952) that this might be the main means by which the chemical information contained in the genes was converted into a geometrical form. If this be so, then any particular type of anatomical structure can only arise from a relatively small fraction of the possible chemical reaction systems: if it arose from all then no other anatomical forms would be possible. In particular, not all reaction systems can be appropriate for the description of phyllotactic phenomena. It is the purpose of this second part of the paper to describe conditions under which the chemical reactions are appropriate for that purpose, to find a partial differential equation which describes the progress of those reactions, and to investigate the behaviour of that partial differential equation. It is unfortunately not possible to say *a priori* how great is the variety of behaviour which could be described by the other reaction systems. It seems to be capable of giving rise to an enormous variety of possible solutions, but whether this variety is, or is not, sufficient to describe all the variety of living forms cannot easily be settled.

The principal assumptions to be made are:

(a) The reaction system is such that there is a homogeneous equilibrium, and small deviations from this equilibrium satisfy the conditions for stationary waves (see TURING (1952) p. 52).

(b) The deviations from the homogeneous state are not very great. They are sufficiently large for the linear approximation to be inapplicable, but nevertheless sufficiently small for the effects of the quadratic and higher terms to be regarded as little more than perturbations.

(c) The only wavelengths which are significant are those which are either very long or fairly near to the optimum.

When it is not intended to assume that the reaction rate functions are linear one may

(i) make some quite definite assumptions about the chemistry of the system and obtain definite reaction rate functions, or,

(ii) admit that the reaction rate functions may be any functions of the concentrations, or,

(iii) assume that the reaction rates are polynomials in the concentrations, or,

(iv) assume that the reaction rates are quadratic functions of the concentrations.

The assumption that the reaction rates are polynomials can be directly justified by the law of mass action. However it gives one very little advantage to know this. Moreover, if one is trying to deal with a particular case one will often prefer to use functions which are not polynomials, since by doing so one can often, if some of the reactions proceed quickly, reduce the number of morphogens, i.e. the number of variables which need to be considered. The assumption that the reaction rates are quadratic functions of the concentrations is a more useful one, although superficially it appears to be rather arbitrary and of limited validity. It can, however, be justified quite as well as the assumption of polynomial reaction rates. The justifying argument may be expressed either in a purely mathematical form or in physical terms, although it is really essentially the same argument in either case. The physical argument is that all chemical reactions in dilute solution are either monomolecular or bimolecular. Where three molecules react there will be two of them which meet and form some, probably more or less unstable, combination, before reacting with the third substance. When these unstable substances are all included as morphogens, there will only be monomolecular and bimolecular reactions, and consequently only linear and quadratic terms in the reaction rates. It may be necessary, if this point of view is to be maintained, to treat excited states of molecules as being molecules of a different compound, but there is clearly no objection to this.

The mathematical argument also proceeds by introducing new variables, and these variables might, if one wished, be imagined as representing the concentrations of imaginary intermediate products according to some possible theory of how the reactions are to be broken down into monomolecular and bimolecular reactions.

The additional variables which are necessary if one is to have quadratic rate functions are clearly no disadvantage in the sort of theoretical discussion where one is in any case committed to an indefinite but finite number of concentration variables.

Although the equations are no longer linear it is convenient to use the same substitutions as in the linear theory, viz. first

$$\Gamma_{mj} = h_m + \sum_r \Gamma_{mj}^{(3)} v_j^{-1/2} l_{rj} \qquad (\text{II}.2.1)$$

where $f_m(h_1, \ldots, h_M) = 0$ for all m, i.e. $h_1, \ldots, h_M$ represents a homogeneous equilibrium. Naturally, if there are variables for every substance which occurs (in solution) in the system, then $h_1 = h_2 = \cdots = h_M = 0$ will be such an equilibrium, but it is preferable for a number of reasons not to suppose that $h_1, \ldots, h_M$ have these values. If some of the substances whose concentrations are effectively constant have not been assigned variables, the values zero may not be an equilibrium. Moreover, the zero equilibrium is by no means a representative one, and not all the phenomena described in TURING (1952) will occur with this trivial form of equilibrium.

The M different roots of the equation resulting from eliminating the $\Gamma^{(3)}_{mr}(0)$ from (II.1.10) may be called $p_l(\alpha)$, $l = 1, 2, \ldots, M$, and there will be numbers $W_{ml}(\alpha)$ such that

$$(p_l(\alpha) + \mu_m \alpha) W_{ml}(\alpha) = \sum_s a_{ms} W_{ms}(\alpha). \tag{II.2.2}$$

If p_l is a k-fold repeated latent root, there will be k corresponding $W_{ml}(\alpha)$. The matrix $W_{ms}(\alpha)$ is accordingly always non-singular and has an inverse $W^{-1}_{st}(\alpha)$. Consequently the matrix $\Gamma^{(3)}_{mr}(t)$ may be expressed in the form

$$\Gamma^{(3)}_{mr}(t) = \sum_l W_{ml}(\alpha_r) X_{lr}(t) \tag{II.2.3}$$

and so, by (II.2.1)

$$\Gamma_{mj}(t) - h_m = \sum_{r,l} W_{ml}(\alpha_r) X_{lr}(t) v_j^{-1/2} l_{rj}. \tag{II.2.4}$$

The assumption (c) above can now be brought into play. Only terms for which α is close to zero or to k_0^2, where $2\pi/k_0$ is the optimum wavelength, must be considered. It is assumed also, that in view of the analogy of the linear case, and assumption (ii) that the non-linear terms are only of secondary importance, one may ignore all terms on the right-hand side of (II.2.4) which do not arise from the largest $\text{Re}\{p_l(\alpha_r)\}$ for the α_r in question. For α_r near to zero this will be supposed to be $p_l(0)(\alpha_r)$, and when α_r is near to k_0^2 it will be supposed to be $l^{(1)}$.

With these assumptions, (II.2.4) becomes

$$\Gamma_{mj}(t) - h_m = \sum_r (\alpha_r \text{ near } 0) W_{ml^{(0)}}(\alpha_r) X_{l^{(0)}r}(t) v_j^{-1/2} l_{rj}$$

$$+ \sum_r (\alpha_r \text{ near } k_0^2) W_{ml^{(1)}}(\alpha_r) X_{l^{(1)}r}(t) v_j^{-1/2} l_{rj}. \tag{II.2.5}$$

A further, and rather drastic, approximation is the assumption that over the two short ranges of values of α concerned, the functions $W_{ml^{(0)}}(\alpha)$

and $W_{ml^{(1)}}(\alpha)$ may be treated as constants, and so

$$\Gamma_{mj}(t) - h_m = W_{ml^{(0)}}(0) \sum_r X_{l^{(0)}r}(t)v_j^{-1/2}l_{rj}$$
$$+ W_{ml^{(1)}}(k_0^2) \sum_r X_{l^{(1)}r}(t)v_j^{-1/2}l_{rj}$$
$$= W_{ml^{(0)}}(0)V_j + W_{ml^{(1)}}(k_0^2)U_j. \tag{II.2.6}$$

Now assuming, in agreement with (iv), that the reaction rates are quadratic functions of the concentrations, the differential equations controlling the behaviour of $\Gamma_{mj}(t)$ can be written in the form

$$\frac{d\Gamma_{mj}}{dt} = \sum a_{ms}(\Gamma_{ms} - h_m) + \sum K_{mrs}(\Gamma_{rj} - h_r)(\Gamma_{sj} - h_s)$$
$$- \frac{\mu_m}{v_j} \sum g_{jj}\Gamma_{mj} \tag{II.2.7}$$

the constant terms being absent by the definition of the h_r. If these equations are expressed in terms of the variables $X_{lr}(t)$ by means of the relation (II.2.4), one obtains

$$\frac{dX_{lr}}{dt} = \sum_{m,j} W_{lm}^{-1}(\alpha_r)l_{rj}v_j^{-1/2}\frac{d\Gamma_{mj}}{dt}. \tag{II.2.8}$$

The linear terms will of course be the same as in the purely linear theory, and the quadratic terms may be evaluated by the approximation (II.2.6). There is evidently no need to give the equations except where $l = l^{(0)}$ or $l = l^{(1)}$. The equation for $X_{l^{(0)}r}$ is

$$\frac{dX_{l^{(0)}r}}{dt} = p_{l^{(0)}}(\alpha_r)X_{l^{(0)}r}$$
$$+ \sum_{m,j,s} W_{l^{(0)}m}^{-1}l_{rj}v_j^{-1/2}K_{mrs}[W_{rl^{(0)}}(0)V_j + W_{rl^{(1)}}(k_0^2)U_j]$$
$$\times [W_{sl^{(0)}}(0)V_j + W_{sl^{(1)}}(k_0^2)U_j] \tag{II.2.9}$$

and may be written

$$\frac{dX_{l^{(0)}r}}{dt} = p_{l^{(0)}}(\alpha_r)X_{l^{(0)}r}$$
$$+ \sum_{m,j} W_{l^{(0)}m}^{-1}l_{rj}v_j^{-1/2}(F_m^{(1)}V_j^2 + 2F_m^{(2)}V_jU_j + F_m^{(3)}U_j^2). \tag{II.2.10}$$

There is of course a similar equation for $X_{l^{(1)}r}$. One can now change the variables to U_j, V_j, and one obtains

[[96]]

$$\frac{dV_j}{dt} = [p_{l^{(0)}}(-\nabla^2)V]_j$$

$$+ \sum W_{l^{(0)}m}^{-1}(\alpha_r)l_{rk}v_k^{-1/2}v_j^{1/2}l_{rj}(F_m^{(1)}V_j^2 + 2F_m^{(2)}V_jU_j + F_m^{(3)}U_j^2). \quad \text{(II.2.11)}$$

Here $p(-\nabla^2)$ represents a certain linear operator. As implied by the notation used, the characteristic vectors of these operators are the same as those of the operator ∇^2, but where $-\nabla^2$ has the characteristic value α, $p(-\nabla^2)$ is to have the characteristic value $p(\alpha)$. If now the matrix $W_{lm}^{-1}(\alpha)$ is treated as independent of α over the range involved, as has already been supposed for its inverse, the equation reduces to the much simplified form

$$\frac{dV_j}{dt} = [p_{l^{(0)}}(-\nabla^2)V]_j + F^{(4)}V_j^2 + 2F^{(5)}V_jU_j + F^{(6)}U_j^2. \quad \text{(II.12.a)}$$

Likewise the equation for U_j takes the similar form

$$\frac{dU_j}{dt} = [p_{l^{(1)}}(-\nabla^2)U]_j + G^{(4)}V_j^2 + 2G^{(5)}V_jU_j + G^{(6)}U_j^2. \quad \text{(II.12.b)}$$

The assumptions which led to these equations have been somewhat drastic, but it must be remembered that it is by no means claimed that these approximations are appropriate for all problems of morphogenesis, but merely that, in the cases where these approximations hold good one obtains equations suitable for the description of the phyllotactic phenomena. When considering the validity of the approximations, therefore, one should ask whether there are any cases in which they hold good, rather than whether they are universally valid.

There will be still further assumptions to be made, but the equations (II.2.12) may be regarded as a convenient bridge linking the more or less chemical theory with the rest of the problem.

A number of additional assumptions will be made with even less appearance of generality. First, the coefficients $F^{(4)}, F^{(5)}, G^{(4)}$ will be assumed to be zero. One may, if one wishes, explain the ignoring of the term $G^{(4)}V_j^2$ on the grounds that, in a Euclidean space at any rate, it does not contain components near to the optimum wavelength; and the ignoring of $2F^{(5)}U_jV_j$ on the grounds that it does not contain terms of long wavelength. The ignoring of $F^{(4)}V_j^2$ might be justified by the view that V_j is small. But whatever justification be offered, these assumptions are made. It is assumed further that there is effective equilibrium in the equation (II.2.12a), i.e. that the right-hand side of this equation vanishes, so that

$$V = [p_{l^{(0)}}(-\nabla^2)]^{-1}U^2 = \psi(-\nabla^2)U^2 \quad \text{(II.2.13)}$$

putting $F^{(6)} = -1$ by suitable scaling of the variables.

The equation for U may be written

$$\frac{dU_j}{dt} = [\phi(-\nabla^2)U]_j GU_j^2 - HU_j V_j \qquad (\text{II.2.14})$$

The two functions ϕ, ψ are bound in theory to be algebraic, but this is of course no real restriction. Any analytic function can be approximated as closely as one pleases by an algebraic function. The essential point about the function $\phi(\alpha)$ is that it has a maximum near $\alpha = k_0^2$. Since it has been supposed in any case that it is only the components of U with wavelengths near to $2\pi/k_0$ which are significant, it can only be the values of $\phi(\alpha)$ for values of α near to k_0^2 which matter, i.e. only the values near to the maximum. An appropriate approximation for $\phi(\alpha)$ therefore seems to be $I(\alpha - \alpha_0)^2$. As regards the function $\psi(\alpha)$, the most natural assumption seems to be that $p_{l^{(0)}}(\alpha)$ increases linearly with increasing α and so that $\psi(\alpha)$ is of the form $A/(B + \alpha)$. However, since computations are greatly simplified if $\psi(\alpha)$ can be put equal to zero for the majority of values of α, other forms for this function will sometimes be used.

3. *Equations for small organisms*

If the dimensions of an organism are not too large in comparison with the optimum wavelength, the characteristic values α_r will usually be well spaced out, except where they are multiple. In this case some of the approximations of the last section will be rather more convincingly justifiable, for there may be only one value of α which needs to be considered in each of the two important ranges, viz. that near the optimum wavelength and that near zero. The value near the optimum wavelength will probably be multiple, but, as has already been mentioned, for a connected organism the root zero is simple: $V_j(t)$ is therefore independent of position. According to the point of view in which V represents the concentration of a diffusing poison, the organism is sufficiently small that the poison may be assumed to be uniformly distributed over it. The function $U_j(t)$, on the other hand, must be a linear combination of diffusion eigenfunctions all with the same eigenvalue, or, in other words, waves with the same wavelength. If $\mathscr{F}$ is a linear operator which removes all components which have not the appropriate wavelength, then the equations may be written

$$\frac{dU_j}{dt} = (P - HV)U_j + G[\mathscr{F}U^2]_j, \qquad (\text{II.3.1})$$

$$V = \overline{U^2}. \qquad (\text{II.3.2})$$

Here P is the value of $p_{f^{(1)}}(\alpha)$ for the particular α (near the optimum) which is concerned. The equilibrium solutions of these equations take the form of multiples of solutions of the equations $U = \mathscr{F}(U^2)$.

This problem may be illustrated by the case where the organism forms a spherical shell. It is not of course possible to build up a spherical shell out of a large number of similarly shaped areas. But if it be built up of a large number of cells which are not quite the same, the effect of the irregularities will become small as the number of cells increases. In any case it will be assumed that the "∇^2" for the shell is the ordinary three-dimensional Laplacian in spherical polar coordinates, with the radial term omitted, and so has the spherical harmonics as characteristic functions. The operator $\mathscr{F}$ will then be one which removes from a function on a sphere all spherical harmonic components except those of a particular degree. To solve the equation $U = \mathscr{F}(U^2)$ is to find a spherical harmonic of that degree which, when it is squared and again has the appropriate orders removed, is unchanged. For each degree there is only a finite number of essentially different solutions of this problem, i.e. solutions which are inequivalent under rotations of the sphere. They have been investigated by B. Richards, and the results are described in Part III.

4. *The equations applied to a plane*

In the case that the cells form an isotropic homogeneous plane (apart from local variations over regions containing not very many cells), the diffusion characteristic functions are plane waves of the form $e^{i(Xx + Yy)}$. Over any finite area of the plane the functions may be approximated as closely as one pleases by Fourier sums

$$\sum C_n e^{i(X_n x + Y_n y)}. \tag{II.4.1}$$

When a function has two independent periods it may be accurately represented by such a series. As was seen in Part I, the exponents (X_n, Y_n) then form a lattice. Such a series can also represent any function accurately within a bounded region, by choosing sufficiently large periods. Functions in the plane can also be approximated by Fourier integrals, and sometimes accurately expressed by them. There is no need, however, to go into these problems of analysis here. There are no organisms consisting of infinite planes of tissue, and it is only of value to consider such imaginary organisms for the light that they throw on other systems more nearly represented in nature. This light will not be appreciably dimmed by making the assumption that the functions concerned have two independent periods, possibly rather large, and the expression (II.4.1) will therefore be supposed exact.

Before going on to the non-linear theory of the plane it will be worth while to ask what sorts of patterns one would get if the linear theory applied throughout. The value of his question lies in the fact that it has a fairly definite answer, and is valuable as an illustration of certain points. Whether the patterns which arise from it can fairly be claimed to occur in nature is another matter.

The equations for the linear case can be obtained by omitting the quadratic terms in (II.2.12b) and so become

$$\frac{dU}{dt} = p_{l^{(1)}}(-\nabla^2)U. \tag{II.4.2}$$

This equation may also be inferred fairly directly from (II.2.9) and (II.2.10) by ignoring all but the largest roots p. The equation (II.4.2) can only be regarded as valid over a limited range of time. Apart from the question (which is being intentionally ignored) of the effects of quadratic terms as t increases, there is the effect of Brownian movement and similar "noise" effects when t has large negative values and U is consequently very small. If this Brownian movement is taken into account, the character of the problem changes. One is no longer trying to find the totality of solutions, or the time development of a solution, but rather to find statistical information about the "ensemble" of solutions.

5. *"Noise" effects*

The equation (II.1.8) can be regarded as accurate only so long as one assumes that Avogadro's number is infinite. Otherwise it will be necessary to admit that the concentration of a morphogen in a cell can only have discrete values, corresponding to the various numbers of molecules that may occur in it. It will also not be possible to predict the actual new concentration at any future time, but only to give probabilities. In some applications of the theory it may be important to consider seriously the possibility that there may be only one or two molecules present, or even none. This would apply, for instance, in the case that the same theory is applied to the spread of epidemics, the "molecules" now being infected and uninfected men, rats, corpses, fleas, etc. It may even apply to the morphogenetic problem for some of the morphogens, e.g. the genes themselves. However, since these statistical effects are in any case of rather secondary importance in this problem it is appropriate to make some simplifying assumptions. It will be supposed in fact that over intervals of time short enough for the concentrations not to undergo appreciable proportionate changes, the number of molecules undergoing any one of the reactions in any one of the cells, or passing through any of the membranes, is large enough that one is justified in using a normal distribution instead of a Poisson distribution for it. Let Γ_{mr} be expressed in gram molecules per unit volume and N be Avogadro's number, then ... [the manuscript breaks off here]

6. *Effects of random disturbances*

If there is one value $p_k(\alpha_r)$ which exceeds all the others, then for almost all initial values the term $\exp\{ip_k(\alpha_r)t\}$ will eventually be far greater than any of the other terms which contribute to Γ_{ms}. In this case then the ultimate condition of the system is almost independent of the initial conditions. Even if there are non-linear terms which eventually have to be taken into account, this will quite possibly not apply until this dominant term has outdistanced the others. However it is not at all uncommon for there to be several different characteristic vectors for which the corresponding diffusion characteristic values α_r are all equal. In this case the "argument by outdistancing" merely tells us that ultimately the only characteristic vectors which need be included in the sum are those for which the $p_k(\alpha_r)$ has its maximum value. It does not, however, say anything about the proportions in which those characteristic vectors are to be taken. According to the theory expounded in §II.1, this could be settled by giving all the concentrations of all the morphogens at some early time. These would then determine the concentrations at all later times. The theory of §II.1 is, however, in error about this. For there are some small effects which are ignored. This does not matter under most circumstances, but sometimes it will, as for instance when the system is near to an unstable equilibrium. The main effects concerned are probably

(1) The statistical nature of the chemical reactions.

(2) The statistical nature of the diffusion.

(3) Variations of the reaction rates from cell to cell, due for instance to the presence of different concentrations of indiffusible catalysts (not reckoned as morphogens).

(4) The irregularity of the cell pattern in the cases where this cell pattern has been idealised in some way, e.g. where it is regarded as forming a spherical surface.

It is pointless to attempt to give any complete list of the effects ignored because of the very approximate nature of the whole investigation, but the above list may be used to illustrate the nature of these effects.

So long as one is interested in the manner of departure from homogeneous equilibrium, one can fairly say that even if non-linear effects are eventually to be considered, nevertheless during the period when these minor disturbances are of significance, the non-linear terms can be ignored. It can happen that a non-homogeneous equilibrium, for whose maintenance the non-linear terms are of importance, may become unstable, and then again these small effects will determine the course that the system will actually take. In these cases, however, there are usually very few alternative courses which can be taken. There is also usually a symmetry which shows that the alternative possible courses are all equally probable: in fact no detailed theory is necessary. In the present section therefore it will be supposed that the linear theory applies. This really means that it is being assumed that the linear theory applies whilst the disturbances operate, but the progress of the differential equation will not be followed up to the time when the non-linear terms become effective. It will also be assumed in this investigation that the multiplicity of the roots of the diffusion characteristic equation is entirely due to symmetry. The situation is very similar to that which arises in the theory of spectra, where the multiplicity of an energy level is almost always due to symmetry.

It seems probable that effects (3) and (4) would normally be of somewhat greater magnitude than the atomistic effects (1) and (2). The atomistic effects have, however, not been ignored: this is mainly because the theory of them is much more satisfactory.

One may, if one wishes, regard a morphogenetic system as described by the MN values of the concentrations of the various morphogens in the various cells, without attempting to classify these variables. If $y_1, \ldots, y_{MN}$ are these variables, the equations may be written in the form

$$\frac{dy_r}{dt} = F_r(y_1, \ldots, y_{MN}, t) + N_r(t), \qquad \text{(II.6.1)}$$

where the functions $F_r(y_1, \ldots, y_{MN}, t)$ describe both the diffusion and the reactions, and $N_r(t)$ describe the statistical effects*. The choice of the values of these functions is to be imagined as independently chosen at different times, although the various values at a particular time may be statistically or even functionally dependent.

---

* The use of these $N_r(t)$ is strictly speaking quite unjustifiable; they could not themselves be given finite values at each time, though their integrals can, and also the integrals below in which they appear.

The value of the integral of $N_r(t)$ over any period of time being a sum of a large number of statistically independent components, arising from any way in which the interval of integration may be broken up into subintervals, has a normal distribution. Its variance is expressible in the form

$$\int_a^b \sigma_r(t) \, dt$$

where (a, b) is the time interval concerned. Likewise, if $g(t)$ is a function statistically independent of $N_r(t)$, then

$$\int_a^b N_r(t)g(t) \, dt$$

is also normally distributed and its variance is

$$\int_a^b \sigma_r(t)[g(t)]^2 \, dt.$$

Now suppose that $y_1(t), \ldots, y_{MN}(t)$ would be an equilibrium at time t if only the conditions were not changing, i.e. that

$$F_r(y_1^{(0)}(t), \ldots, y_{MN}^{(0)}(t), t) = 0 \qquad (\text{II.6.2})$$

and let the partial derivative $\partial F_r/\partial y_s$ have the value $b_{rs}(t)$ for the arguments $y_1^{(0)}(t), \ldots, y_{MN}^{(0)}(t), t$. Then putting $y_r^{(1)}(t) = y_r(t) - y_r^{(0)}(t)$ one has

$$\frac{dy_r^{(1)}}{dt} = \sum_s b_{rs}(t) y_s^{(1)}(t) + N_r(t) - \dot{y}_r^{(0)}(t) \qquad (\text{II.6.3})$$

with a dot denoting differentiation with respect to t. Similar equations hold if the variables are subjected to a linear transformation. The solution of the equation may be expressed in the form

$$y_r^{(1)}(t) = \sum_s \int_{-\infty}^t (N_s(u) - \dot{y}_s^{(0)}(u)) K_{rs}(t, u) \, du \qquad (\text{II.6.4})$$

where

$$\frac{\partial K_{rs}}{\partial t} = \sum_q b_{rq}(t) K_{qs}(t) \quad \text{if } t > u \qquad (\text{II.6.5})$$

and

$$K_{rs}(u, u) = \delta_{rs}. \qquad (\text{II.6.6})$$

One may regard $b_{rs}(t)$ as differing rather little from some "ideal" coefficients $b_{rs}^{(0)}(t)$ which have the appropriate symmetry. Similarly, one may have $K_{rs}^{(0)}$ related to $b_{rs}^{(0)}$ as in (II.6.5). In the case where the b_{rs} are independent of time, the $K_{rs}(t)$ are sums of time exponentials with the $p_k(\alpha_r)$ as growth rates. These $p_k(\alpha_r)$ are of course the characteristic values of the matrix b_{rq}. The small difference between b_{rs} and $b_{rs}^{(0)}$ will make only a small difference in the characteristic values, but if the time lapse is large the effect on the corresponding exponential growth factor may be considerable. With the actual $b_{rs}(t)$ which may be encountered, which are not constant in time, the same considerations of magnitude will apply.

It will not be feasible to give a very complete account of the behaviour of the solutions, but some general conditions may be considered. The case in which there are only atomistic disturbances, i.e. of kinds (1) and (2), can be dealt with fairly fully. In this case the term

$$-\int_{-\infty}^{t} \ddot{y}_s^{(0)} K_{rs}(t, u)\, du$$

in (II.6.4) represents an effect independent of position. That this is so is not at all obvious from the equation, which does not in any case distinguish space and morphogen effects. It may be seen from the general principle that in the absence of any of the effects (1) to (4) a solution homogeneous at one time will remain so thereafter. Concentrating on the space dependent term

$$\int_{-\infty}^{t} N_s(u) K_{rs}(t, u)\, du = y_r^{(2)}(t)$$

one sees at once that each of its components is normally distributed, and that every linear combination of components is also normally distributed. The vector $y_1^{(2)}, \ldots, y_{MN}^{(2)}$ consequently has a frequency function of the form

$$(\varDelta(t))^{-1/2}(2\pi)^{-MN/2} \exp\left[-\frac{1}{2}\sum_{i,j} \theta_{ij}(t)\, y_i^{(2)} y_j^{(2)}\right] \qquad (\text{II.6.7})$$

where the quadratic form $\sum \theta_{ij}(t)\, y_i^{(2)} y_j^{(2)}$ is positive definite and $\varDelta(t)$ is the determinant of its coefficients.

It is proposed to consider this matrix θ_{ij} in respect of its asymptotic behaviour in time, and the symmetry of the system. For any given t, the function $K_{rs}(t, u)$ will normally be small compared with its maximum except over a comparatively narrow range of values of u, viz. near to the point where the largest real part of a characteristic value of the matrix $b_{rs}(u)$ is zero (cf. TURING (1952) equation (9.15)). For large values of t, therefore, one may replace the lower limit of integration in (II.6.4) by some fixed t_0, and then $y_r^{(2)}(t)$ will satisfy the differential equation

$$\frac{dy_r^{(2)}}{dt} = \sum_s b_{rs}(t)\, y_s^{(2)}$$

and consequently the matrix θ_{ij} will satisfy

$$\frac{\mathrm{d}\theta_{ij}}{\mathrm{d}t} = -\sum_p (\theta_{ip} b_{jp} - \theta_{jp} b_{ip}). \tag{II.6.8}$$

If one makes a linear transformation to variables $y_r^{(3)}(t)$ such that the corresponding matrix $b_{rs}^{(3)}(t)$ is diagonal, these variables will be, or could be, the same as the $\Gamma_{rs}^{(3)}$, and the characteristic values $b_{ii}^{(3)}$ are the $p_k(\alpha_r)$. Writing I for the largest of the real parts of these characteristic values and $y_r^{(4)}(t)$ for $y_r^{(3)}(t) \, \mathrm{e}^{-It}$, then $y_r^{(4)}(t)$ tends to a limit $y_r^{(5)}$ as t tends to infinity. This limit is zero unless $\mathrm{Re}(b_{rr}^{(3)} - I) = 0$, and when this happens the corresponding imaginary part $\mathrm{Im}(b_{rr}^{(3)})$ is zero, since it is assumed that the stationary wave case applies. The differential equation (II.6.8) becomes, when applied to the coefficients $\theta_{ij}^{(4)}$ which describe the distribution of $y_r^{(4)}$,

$$\frac{\mathrm{d}\theta_{ii}^{(4)}}{\mathrm{d}t} = -2(b_{ii}^{(3)} - I)\theta_{ii}$$

($\theta_{ij} = 0$ if $i \neq j$), and so θ_{ii} tends to infinity unless $b_{ii}^{(3)} = I$. This merely expresses that it is infinitely improbable that $y_r^{(5)}$ will not be zero if $b_{rr}^{(5)} = I$. The $\theta_{rr}^{(4)}$ with $b_{rr}^{(3)} = I$ give the frequency distribution of the $y_r^{(5)}$. They are normally distributed and independent.

The above argument gives an existence theorem about the distribution of the $y_r^{(5)}$, i.e. of those linear combinations of the coordinates which tend to infinity fastest. The actual distribution is best obtained by symmetry requirements. Although the full details of these arguments must be related to the particular symmetry group involved, a little can be said which is generally applicable.

It has already been mentioned that the degeneracy of the characteristic equation associated with diffusion was to be supposed entirely due to symmetry. This assumption needs some clarification before it can be used. Let y be any vector (i.e. an assignment of morphogen concentrations to the cells), then this vector can be transformed into various others, e.g. $S_p y$ by the symmetry operations, i.e. the permutations τ of the cells which satisfy

$$g_{p_r p_s} = g_{rs}.$$

If y is a characteristic vector of the operator ∇^2 with characteristic value α, then $S_p y$ is also a characteristic vector with the same characteristic value. The various vectors

$$S_{p_1} y, S_{p_2} y, \ldots$$

span a space of vectors with characteristic value α. In the metaphor of the assumption on p. 102, the equality of these characteristic values is "due to the symmetry''. That assumption states that the characteristic values are equal only if equality is due to symmetry, i.e. if y, y' are both characteristic vectors with characteristic value α then y' can be expressed in the form $\sum \beta_p S_p y$. Another way of expressing the condition is to say that when the group of symmetry is represented on the vectors with characteristic value α, this representation is irreducible. What is actually required below is a little more, viz. that the degeneracy of the largest roots of the matrix $b_{rs}(t)$ (for large t) should be entirely due to symmetry. These roots are the $p_k(\alpha_r)$. The degeneracies of the diffusion matrix mean that various sets of the α_r are equal. It is required that if $I = \text{Max}[p_k(\alpha_r)]$ then $p_k(\alpha_r) = I$ and $p_{k'}(\alpha_{r'}) = I$ imply that $\alpha_r = \alpha_{r'}$ and $k = k'$. The cases in which this is most likely to be incorrect are when either there is some kind of symmetry in the chemical reaction system, as for instance when the dextro and laevo forms of the morphogens are distinguished, or when $\alpha_r \neq \alpha_{r'}$, $k = k'$, but $p_k(\alpha_r) = p_k(\alpha_{r'})$ due to fortuitous values of the geometrical dimensions. With these assumptions, the representation of the symmetry group on the vectors for which b_{rs} has characteristic value I is irreducible. Now let the values of r for which $b_{rr}^{(3)} = I$ be $1, 2, \ldots, J$, and let the effect of the symmetry operation on $y_1^{(5)}, \ldots, y_5^{(5)}, 0, 0, \ldots, 0$ be to convert it to the vector whose rth coordinate $(r \leqslant J)$ is

$$\sum_{s=1}^{J} u_{rs}(p) y_s^{(5)}.$$

Also let the frequency distribution of the values of $y_1^{(5)}, \ldots, y_J^{(5)}$ be

$$(2\pi)^{-J/2} (\det \theta^{(5)})^{-J/2} \exp\left[-\frac{1}{2} \sum_{i,j} \theta_{ij}^{(5)} y_i^{(5)} y_j^{(5)} \right] dy_1^{(5)}, \ldots, dy_J^{(5)}.$$

The frequency distribution for the transformed coordinates will have

$$\sum_{k,l} u_{ik}(\tau^{-1}) u_{jl}(\tau^{-1}) \theta_{kl}$$

in place of $\theta_{ij}^{(5)}$, but by the symmetry condition the frequency distribution is unaltered, i.e. $\sum u_{ik}(\tau^{-1}) u_{jl}(\tau^{-1}) \theta_{kl} = \theta_{ij}^{(5)}$.

Now for a finite group of symmetry, the basis of the vectors may be chosen in such a way that the matrices of the representation are unitary. If this is done, the condition may be expressed by saying that the matrix θ_{ij} commutes with each representative matrix $u_{ij}(\tau^{-1})$. But by a well known theorem on representations, this means that the matrix θ_{ij} is a multiple of the unit matrix.

When the coordinates are chosen in the space of vectors of greatest growth rate in such a way that the representatives of the symmetry group are unitary, then the frequency distribution is normal in each coordinate and the coordinates are independent of one another and have equal variances.

The consequences of the non-atomistic effects (3) and (4) are not capable of so satisfactory a treatment. If one were justified in supposing that the final effects were linear functions of independent causes the outcome of effects (3) and (4) would be of the same kind as the outcome of effects (1) and (2). It is likely however that this is not usually a good approximation. Although it would be fair to regard the matrix $(b_{rs}) = B$ as influenced by a number of linearly independent causes, this matrix takes effect mainly through an exponential $e^{B\tau}$ with a fairly large value of τ. Unless $\varepsilon\tau$ is small, $e^{(B+\varepsilon B')\tau}$ cannot be satisfactorily approximated by a linear function of B'.

It is not proposed to enter into these effects here in great detail, although it is probable that they are of greater importance than those of atomistic origin. It is only intended to deal with an extreme case, one which is very different from the case of the atomistic effects, viz. that in which the time period is so long that in the exponential $e^{(B+\varepsilon B')\tau}$ the differences of the characteristic values of the matrix, when multiplied by the time interval, are large. In this case one comes back to the situation in §II.1, where only the largest characteristic value need be considered. The actual characteristic value is of no particular interest, as it will merely determine the time before a pattern of a given amplitude appears. Greater interest attaches to the characteristic vector, as it determines the pattern.

Part III. A Solution of the Morphogenetical Equations for the Case of Spherical Symmetry

1. *Reduction of the differential equation*

In the previous part the equations for small organisms were deduced, and an attempt was made to justify the assumptions on a chemical basis. One is interested in finding the concentration function $U(\theta, \phi, t)$ satisfying the equations:

$$\frac{dU}{dt} = \Phi(\nabla^2)U + GU^2 - HUV, \qquad \text{(III.1.1)}$$

$$V = \overline{U_2}. \qquad \text{(III.1.2)}$$

It has been shown that the solution U must be a linear combination of diffusion eigenfunctions; it is therefore appropriate to assume that the solution of (III.1.1) is of the form

$$U(\theta, \phi, t) = \sum_{m=-n}^{m=n} S_m(t) \overline{P_n^m}(\cos \theta) \, e^{im\phi} \qquad \text{(III.1.3)}$$

with U real and the functions $\overline{P_n^m}(\cos \theta)$ being the normalised Legendre Associated Functions (see Appendix). These functions satisfy the relation

$$\frac{1}{4\pi} \int \int \overline{P_n^r}(\cos \theta) \overline{P_n^s}(\cos \theta) \, \mathrm{d}S = \begin{cases} 1 & \text{for } r=s, \\ 0 & \text{for } r \neq s. \end{cases} \qquad \text{(III.1.4)}$$

Since U is real, it is equal to its complex conjugate U^*, and since

$$\overline{P_n^m}(\cos \theta) = \overline{P_n^{-m}}(\cos \theta)$$

it follows that

$$S_{-m}(t) = S_m^*(t). \qquad \text{(III.1.5)}$$

It is then possible to evaluate V from (III.1.2) by integration, since

$$\begin{aligned} V &= \frac{1}{4\pi} \int \int U^2 \, \mathrm{d}S \\ &= \frac{1}{4\pi} \int \int \sum_{k=-n}^{k=n} \sum_{m=-n}^{m=n} S_k S_m \overline{P_n^k}(\cos \theta) \overline{P_n^m}(\cos \theta) \, e^{i(k+m)\phi} \, \mathrm{d}\cos \theta \, \mathrm{d}\phi \\ &= \sum_{k=-n}^{k=n} \sum_{m=-n}^{m=n} S_k S_m \frac{1}{4\pi} \int \int \overline{P_n^k}(\cos \theta) \overline{P_n^m}(\cos \theta) \, e^{i(k+m)\phi} \, \mathrm{d}\cos \theta \, \mathrm{d}\phi. \end{aligned}$$

Now using the relations (III.1.4) and (III.1.5), it follows that

$$V = \sum_{m=-n}^{m=n} |S_m|^2. \qquad \text{(III.1.6)}$$

In equation (III.1.1), the function $\Phi(\nabla^2)$ is replaced by a constant I, since for the case of spherical surface harmonics,

$$\nabla^2 = -n(n+1)/R^2.$$

Equation (III.1.1) thus becomes:

$$\frac{\mathrm{d}U}{\mathrm{d}t} = IU + GU^2 - HUV. \qquad \text{(III.1.7)}$$

Substituting the series solution into this equation leads to:

$$\frac{\mathrm{d}S_m}{\mathrm{d}t} = (I - HV)S_m + G \sum_{i=-n}^{i=n} \sum_{j=-n}^{j=n} S_i S_j L_n^{i,j,-m} \qquad \text{(III.1.8)}$$

the coefficients $L_n^{p,q,r}$ being defined by:

[108]

$$L_n{}^{p,q,r} = \frac{1}{4\pi} \int \int \overline{P_n{}^p}(\mu) \overline{P_n{}^q}(\mu) \overline{P_n{}^r}(\mu)\, e^{i(p+q+r)\phi}\, d\mu\, d\phi, \quad \text{(III.1.9)}$$

$-\mu$ being written for $\cos\theta$.

As time goes on, $S_m(t)$ will reach its equilibrium value as determined by equation (III.1.8). Therefore, assuming that equilibrium has been reached, this equation becomes

$$(I - HV)S_m + G \sum_{i=-n}^{i=n} \sum_{j=-n}^{j=n} S_i S_j L_n{}^{i,j,-m} = 0 \qquad \text{(III.1.10)}$$

which can be written in the form

$$-\left(\frac{I-HV}{G}\right)S_m = \sum_{i=-n}^{i=n} \sum_{j=-n}^{j=n} S_i S_j L_n{}^{i,j,-m}. \qquad \text{(III.1.11)}$$

For the purposes of obtaining a solution it is more convenient to solve the equation

$$T_m = \sum_{i=-n}^{i=n} \sum_{j=-n}^{j=n} T_i T_j L_n{}^{i,j,-m} \qquad \text{(III.1.12)}$$

which does not contain any of the arbitrary constants. If T_m is a solution of this equation, then $S_m = \lambda T_m$ will satisfy the equation

$$\lambda S_m = \sum_{i=-n}^{i=n} \sum_{j=-n}^{j=n} S_i S_j L_n{}^{i,j,-m} \qquad \text{(III.1.13)}$$

and thus λT_m will satisfy (III.1.11) provided one chooses λ so that

$$\lambda = -(I - HV)/G. \qquad \text{(III.1.14)}$$

Here the constant V is given by

$$V = \sum |S_m|^2 = \lambda^2 \sum |T_m|^2 \qquad \text{(III.1.15)}$$

so that the final equation for λ is

$$\lambda = -\frac{I - \lambda^2 H \sum |T_m|^2}{G}. \qquad \text{(III.1.16)}$$

Since λ is the same for each of the T_m, there will only be a constant factor between the distribution given by the T's and that given by the S's.

One thus arrives at the set of equations

$$S_m = \sum_{i=-n}^{i=n} \sum_{j=-n}^{j=n} S_i S_j L_n{}^{i,j,-m} \qquad \text{(III.1.17)}$$

which are to be solved for the unknown S_m. The solutions are in general complex, but it is possible to choose solutions which are purely real, since it can be shown that the complex solutions differ only by a rotation.

It is possible to enumerate the $(n+1)$ equations as given by (III.1.17) for each particular value of n, in the forms given below. The relations

$$S_{-m} = S_m^*,$$

$$L_n^{r,s,-m} = E_n^{|r|,|s|,|m|} \quad \text{where } r+s-m=0, \qquad \text{(III.1.18)}$$

$$E_n^{p,q,r} = 0 \quad \text{if any of } p,q,r \text{ is } > n$$

have been used.

For $n \leqslant 8$ the equations are

$$S_0 = S_0^2 E_n^{0,0,0} + 2 \sum_1^n |S_k|^2 E_n^{0,0,0},$$

$$S_1 = 2 \sum_0^{n-1} S_k^* S_{k+1} E_n^{1,k,k+1},$$

$$S_2 = 2 \sum_0^{n-2} S_k^* S_{k+2} E_n^{2,k,k+2} + S_1^2 E_n^{1,1,2},$$

$$S_3 = 2 \sum_0^{n-3} S_k^* S_{k+3} E_n^{3,k,k+3} + 2 S_1 S_2 E_n^{1,2,3},$$

$$S_4 = 2 \sum_0^{n-4} S_k^* S_{k+4} E_n^{4,k,k+4} + 2 S_1 S_3 E_n^{1,3,4} + S_2^2 E_n^{2,2,4},$$

$$S_5 = 2 \sum_0^{n-5} S_k^* S_{k+5} E_n^{5,k,k+5} + 2 S_1 S_4 E_n^{1,4,5} + 2 S_2 S_3 E_n^{2,3,5},$$

$$S_6 = 2 \sum_0^{n-6} S_k^* S_{k+6} E_n^{6,k,k+6} + 2 S_1 S_5 E_n^{1,5,6}$$
$$+ 2 S_2 S_4 E_n^{2,4,6} + 2 S_3^2 E_n^{3,3,6},$$

$$S_7 = 2 \sum_0^{n-7} S_k^* S_{k+7} E_n^{7,k,k+7} + 2 S_1 S_6 E_n^{1,6,7}$$
$$+ 2 S_2 S_5 E_n^{2,5,7} + 2 S_3 S_4 E_n^{3,4,7},$$

$$S_8 = 2 \sum_0^{n-8} S_k^* S_{k+8} E_n^{8,k,k+8} + 2 S_1 S_7 E_n^{1,7,8}$$
$$+ 2 S_2 S_6 E_n^{2,6,8} + 2 S_3 S_5 E_n^{3,5,8} + S_4^2 E_n^{4,4,8}$$

where for $n < 8$ only the first $n+1$ equations hold.

These equations have algebraic solutions, though they are not necessarily real. The solutions $S = (S_0, S_1, \ldots, S_n)$ for given n are not unique, and indeed it is sometimes possible to specify n of the $n+1$ components of S and to vary the remaining component, keeping its modulus fixed. Solutions have been sought with $S_1 = 0$, since in any solution with $S_1 \neq 0$ it is possible to rotate the coordinate axes so as to eliminate S_1' in the resulting transformation $S \rightarrow S'$. In doing this, the same physical solution is preserved.

It will be seen that in the solutions obtained, some values are negative. It is to be remembered that the solutions represent deviations from the sphere. The dimensions are somewhat arbitrary, but a correct balance between the oscillations of the function U and the radius of the initial sphere can be obtained by reference to suitable biological species.

2. Solutions of the simultaneous equations

The solutions of the sets of equations conform, in general, to a set pattern. If there is a solution in which S_r is non-zero, then S_{2r}, S_{3r}, etc. will also occur in that particular solution. Thus there will be a solution with S_0, S_3, S_6 etc. all non-zero; or again, S_0, S_2, S_4, S_6 etc. non-zero. There will be additional solutions if certain coincidences in numerical values hold among the integrals $L_n^{i,j,k}$.

The equations are solved for the values of $n = 2, 4, 6$, together with the restriction mentioned previously, namely that $S_1 \equiv 0$.

Case $n = 2$. The equations are:

$$S_0 = S_0^2 E_2^{0,0,0} + 2S_1^* S_1 E_2^{1,1,0} + 2S_2^* S_2 E_2^{2,2,0},$$
$$S_1 = 2S_1 S_0 E_2^{1,1,0} + 2S_1^* S_0 E_2^{1,1,2}, \qquad \text{(III.2.1)}$$
$$S_2 = 2S_2 S_0 E_2^{2,2,0} + S_1^2 E_2^{1,1,2}$$

while the coefficients E_n have numerical values

$$E_2^{0,0,0} = 2\sqrt{5}/7, \qquad E_2^{1,1,0} = \sqrt{5}/7,$$
$$E_2^{2,2,0} = -2\sqrt{5}/7 \qquad E_2^{1,1,2} = \sqrt{30}/7.$$

The simplest solution of this set (III.2.1) is that for which S_0 is the only non-zero variable. This solution is

$$S_0 = (E_2^{0,0,0})^{-1} = 7\sqrt{5}/10$$

and hence

$$U = S_0 \overline{P_2^0}(\cos \theta) = (7/4)(3 \cos^2 \theta - 1),$$
$$V = S_0^2 = 49/20.$$

This gives rise to a prolate spheroid whose major axis coincides with the direction $\theta = 0$. All other solutions of this set of equations do not introduce any physical solutions but only rotate the spheroid through varying angles about different axes. For example, consider the solution with $S_0 \neq 0$, $S_2 \neq 0$. We can choose S_2 to be positive, for the case S_2 negative is obtained by rotation through $\pi/2$ about the axis $\phi = 0$, since

$$-S_2 \overline{P_2^2}(\mu) \cos 2\phi = S_2 \overline{P_2^2} \cos 2(\phi \pm \pi/2).$$

The solution is then

$$U = (7/8)[(1 + 3 \cos 2\phi) - 3\mu^2 (1 + \cos 2\phi)],$$

$$V = 49/20.$$

This is seen to be the same prolate spheroid as the original, but merely rotated through $\pi/2$ about the axis $\phi = \pi/2$.

Case $n = 4$. (i) Here again the simplest solution is a fundamental one physically; the solution being that for which S_0 is the only non-zero coefficient. It is

$$U_r = (E_4^{0,0,0})^{-1} \overline{P_4^0}(\cos \theta) = (1001/1296)(35\cos^4\theta - 30\cos^2\theta + 3),$$

$$V = (1001/486)^2.$$

This is a solid of revolution about the polar axis and resembles a discoid elongated in the polar directions.

In identifying the physical solutions, the value of the integral V, as calculated from (II.3.1), serves as an identity number, since for two physical solutions to be the same they must have the same wavelength predominating and must have the same value for V.

(ii) If one assumes two non-zero coefficients, namely S_0 and S_4, the solution is

$$U = S_0 \overline{P_4^0}(\mu) + 2S_4 \overline{P_4^4}(\mu) \cos 4\phi$$

$$= (143/108)[(3/8)(35\mu^4 - 30\mu^2 + 3) + (30/16)(1 - \mu^2)^2 \cos 4\phi],$$

$$V = 20449/6804.$$

This solid is a spheroid with a spine at each pole and four around the equator; in fact it has six-point symmetry. It is thus totally different from the previous solution.

(iii) The solution with S_0 and S_3 as the only non-zero coefficients yields the same solid, the solution being

$$U = S_0 \overline{P_4^0}(\mu) + 2S_3 \overline{P_4^3}(\mu) \cos 3\phi$$
$$= -(143/324)[(3/4)(35\mu^4 - 30\mu^2 + 3) - 15\sqrt{2}(1 - \mu^2)^{3/2}\mu \cos 3\phi],$$
$$V = 20449/6804.$$

In this case, however, the pole has been rotated through $\cos^{-1}(1/\sqrt{3}) = 54° \, 44'$.

(iv) In seeking a further solution one makes the assumption that $S_0 \neq 0$, $S_2 \neq 0$. If one further assumes that these two are real, one can deduce that S_4 is also real and non-zero. Since the solution of a quadratic in S_0 is required, one expects two solutions; in fact these two solutions are those previously obtained, namely, the discoid and the six-spined spheroid, the former having the line $\phi = \pi/2$ for its axis of rotation, while the latter has been rotated about the axis $\phi = \pi/2$. The solutions are

$$S_0 = -1001/3024, \qquad S_2 = 1001\sqrt{10}/3024,$$
$$S_4 = 1001\sqrt{70}/144\,112, \qquad V = 20449/6804,$$

and

$$S_0 = 1001/1296, \qquad S_2 = 1001\sqrt{90}/11\,664,$$
$$S_4 = 1001\sqrt{70}/7776, \qquad V = (1001/486)^2.$$

(v) The only other solution in the pattern results from the assumption that S_2 is real and S_3 is non-zero. This most general solution is obtained due to a certain numerical relationship among the E's. The value of the integral is

$$V = (1001/486)^2$$

and thus it is the discoid, having, in fact, been rotated about the axis $\phi = \pi/2$ through an angle of $50°$.

For this case, $n = 4$, there are thus only the two physical solutions, the discoid and the six-spined spheroid.

Case $n = 6$. (i) For the case $n = 6$, the solid of revolution is again the easiest solution to obtain and corresponds to the function $\overline{P_6^0}(\cos \theta)$ in that it has two ridges equidistant from the equator. Thus the solution is

$$U = (3553\sqrt{13}/5200)\overline{P_6^0}(\cos \theta),$$
$$V = 13(3553/5200)^2.$$

(ii) A more interesting solution is that for which S_0 and S_5 are non-zero, the solution being

$$U = (323\sqrt{13}/1300)[\overline{P_6^0}(\mu) + (323\sqrt{1001}/14\,300)\overline{P_6^5}(\mu)\cos 5\phi],$$

$$V = 104\,329/57\,200.$$

This corresponds to a regular icosahedron, the twelve equal spines being separated by the correct angular distance of 63°24', this particular solution having spines at the poles.

(iii) Assume S_3, S_5 to be zero and S_2, S_4, S_6 to be real. Now because $E_6^{2,2,0} = E_6^{6,6,0}$ it is possible to find a simpler solution in which $S_4 = 0$ because the equations are compatible. The solution of the reduced equations then leads to

$$U = -\frac{323\sqrt{13}}{520}\left[\overline{P_6^0}(\mu) + \sqrt{\frac{21}{29}}\,\overline{P_6^2}(\mu)\cos 2\phi + \sqrt{\frac{378}{319}}\,\overline{P_6^6}(\mu)\cos 6\phi\right],$$

$$V = \frac{5\,529\,437}{228\,800}.$$

This gives rise to a somewhat irregular solid with ten spines.

(iv) The solution in which S_0 and S_6 are assumed to be non-zero gives a similar solution with the same value of V.

(v) The icosahedron mentioned above appears again in the solution with S_0, S_3, S_6 all non-zero.

The solution with S_0 and S_4 non-zero has twelve equal spines situated about the polar axis as the axis of symmetry, and is really solution (i) slightly modified by the $\cos 4\phi$ term.

3. Comparison with physical species

The biological group which best illustrates the spherical harmonic pattern is that of the Radiolaria. These marine organisms are unicellular, and are surrounded by a skeleton for support and protection, this latter being generally composed of silica. These small Radiolarian cells are about a millimetre in diameter, and are found in all the seas of the world, in all climactic zones, and at all depths, but are not found in fresh water. Their most interesting property from the present point of view is that of possessing radial spines which radiate from the outer shell of the skeleton. From a morphological aspect the number, the arrangement and disposition of the spines is usually the determining factor regarding the general form of the skeleton. Physiologically they discharge distinct functions as organs of protection and support.

The life of a single cell is essentially individual and its growth is influenced by the surroundings. Thus it can be conceived that the numerous forms that abound are due to various concentrations of diffusing materials both organic and inorganic. The salinity of the water or the silica content may be likened to the poison morphogen, as these are known to influence the growth.

For the purposes of biological classification, the Radiolaria are divided into two subclasses, the "Porulosa" and the "Osculosa", and are further subdivided into four legions. The subclass Porulosa includes the two legions SPUMELLARIA and ACANTHARIA, which have the following characteristics:

(1) The central capsule is a sphere and retains this form throughout the majority of the species.

(2) The equilibrium of the floating unicellular body is either pantostatic (indifferent) or polystatic (plural-stable), since a vertical axis is either absent, or if present has its two poles similarly constituted.

(3) The ground forms of the skeleton are therefore almost always spherotypic or isopolar-monaxon, very rarely zygotypic.

The subclass Osculosa comprises the two legions NASSELARIA and PHAEODARIA, which agree in similar and constant characteristics.

(1) The central capsule is a sphere and retains this ground form in most of the species.

(2) The equilibrium of the floating body is monostatic and unistable, since the two poles of the main axis are always more or less different from each other.

(3) The ground forms of the skeleton are, therefore, for the most part grammotypic (centraxon) or zygotypic, rarely spherotypic.

The four principal groups of Radiolaria, which have been given the name "legions", are natural units; when, however, the attempt is made to bring them all into a phylogenetic relationship it undoubtedly appears that the SPUMELLARIA are the primitive stem. The other three have developed, probably independently, from the most ancient stem form of the SPUMELLARIA, the spherical "Actissa".

As our main interest will lie in the development of the individual ground forms, it will be appropriate to give a survey of the various types. We can classify the great variety of the geometrical ground forms into four principal groups: the "Centrostigma" or Spherotypic, the "Centraxonia" or Grammotypic, the "Centroplana" or Zygotypic, and the "Acentrica" or Atypic. The natural centre of the body, about which all its parts are regularly arranged, is in the first group a point (stigma), in the second a straight line (principle axis), in the third a plane (sagittal plane) and in the fourth a centre is, of course, wanting.

The Spherical or Homaxon ground form is the only absolutely regular ground form, since only in it are all axes which pass through the centre equal. It is very often realised among the Radiolaria, especially in the SPUMELLARIA and in the ACANTHARIA, where it furnishes the common original ground form, but it is often to be seen in the shells of many PHAEODARIA. On the other hand, it is never found among the NASSELLARIA.

The endospherical polyhedron or polyaxon ground form naturally follows the spherical or homaxon. Under it are included all polyhedra whose angles fall on the surface of a sphere; this ground form is especially common among the SPUMELLARIA but it is also found among the ACANTHARIA. Strictly speaking, all those lattice shells which have been incorrectly called "spherical" belong to this category, for none of them are true spheres in the geometrical sense, but rather endospherical polyhedra, whose angles are indicated by the nodal points of the lattice shell or the radial spines which spring from them. These polyhedra may be divided into three groups, regular, subregular and irregular. Of the regular polyhedra, properly so-called, only five can exist, namely the regular tetrahedron, cube, dodecahedron, octahedron and icosahedron. All these are actually manifested among the Radiolaria, although the subregular endospherical polyhedra are much more common.

The ground form whose geometrical type is the regular icosahedron (bounded by twenty equilateral triangles) occurs among the PHAEODARIA (e.g. "Circongonia") and also in certain Aulosphaerida. This ground form may also be assumed to occur in those Sphaeroidea whose spherical lattice shells bear twelve equal and equidistant radial spines; the basal points of these spines indicate the twelve angles of the regular icosahedron.

The regular octahedron (bounded by eight equilateral triangles) commonly appears among the SPUMELLARIA. In these Sphaeroidea the typical ground form is usually indicated by six equal radial spines which lie on three perpendicular axes. Occasionally the spherical form of the lattice shell passes over into that of the regular octahedron. The same form recurs in "Circoporus" among the PHAEODARIA.

The regular cubic ground form and the regular tetrahedral ground form also occur. The former may be regarded as occurring in those species whose spherical lattice shell bears eight equal and equidistant radial spines. The isopolar-monaxon ground form is characterised by the possession of a vertical main axis with equal poles, whilst no transverse axes are differentiated. All horizontal planes which cut the axis at right angles are circles. The most important ground forms of this group are the "phacoids" (the lens or oblate spheroid) and the ellipsoid (or prolate spheroid) e.g. *Phacodiscus rotula* and *Cromyactractus tetracelyphus*.

Appendix

The functions $\overline{P_n^m}(\cos\theta)$ are the normalised Legendre Associated Functions defined by

$$\overline{P_n^m}(\mu) = A_n^m P_n^m(\mu),$$

where μ is written for $\cos\theta$ and $P_n^m(\mu)$ represents the usual Legendre Associated Functions, with the condition that $P_n^m(\mu) = P_n^{-m}(\mu)$ and A_n^m is chosen so that

$$\frac{1}{4\pi} \iint_S \overline{P_n}^{m_1}(\mu) \overline{P_n}^{m_2}(\mu)\, e^{i(m_1 + m_2)\phi}\, d\cos\theta\, d\phi = 1.$$

One therefore finds that

$$A_n^m = \sqrt{\frac{(2n+1)(n-m)!}{(n+m)!}}$$

(see Hobson (1931) p. 162).

It is convenient to introduce the functions $L_n^{p,q,r}$ and $E_n^{p,q,r}$, defined by:

$$L_n^{p,q,r} = \frac{1}{4\pi} \int_0^{2\pi} \int_{-1}^1 \overline{P_n^p}(\mu) \overline{P_n^q}(\mu) \overline{P_n^r}(\mu)\, e^{i(p+q+r)\phi}\, d\mu\, d\phi$$

and

$$E_n^{p,q,r} = \tfrac{1}{2} \int_{-1}^1 \overline{P_n^p}(\mu) \overline{P_n^q}(\mu) \overline{P_n^r}(\mu)\, d\mu.$$

It follows that

$$L_n^{p,q,r} = \begin{cases} E_n^{p,q,r} = E_n^{|p|,|q|,|r|} & \text{for } p+q+r=0, \\ 0 & \text{for } p+q+r \neq 0. \end{cases}$$

Since the definition of $L_n^{p,q,r}$ is independent of choice of axis, it is possible to rotate the coordinate axis and keep the integral constant. This enables one to obtain a form of recurrence relation for the L's:

$$m_1 K_n^{m_1-1} L_n^{m_1-1,m_2,m_3} + m_2 K_n^{m_2-1} L_n^{m_1,m_2-1,m_3} + m_3 K_n^{m_3-1} L_n^{m_1,m_2,m_3-1} = 0$$

subject to the condition that $m_1 + m_2 + m_3 = 1$; the constants $^p K_n^q$ being given by

$$^{p+1}K_n{}^p = -\tfrac{1}{2}\sqrt{(n-p)(n+p+1)},$$

$$^pK_n{}^{p+1} = \tfrac{1}{2}\sqrt{(n-p)(n+p+1)},$$

$$^pK_n{}^q \equiv 0 \quad \text{unless } |p-q| = 1,$$

$$^pK_n{}^q = {}^{|p|}K_n{}^{|q|}.$$

The set of equations which arises from the recurrence relation is not unique, and it is therefore possible to derive several checking equations. For the case $n=2$ the set for solution contains three equations since, in all cases, it is necessary to determine $E_n{}^{0,0,0}$ beforehand. This is best done by direct calculation from the integral or by use of the formula:

$$E_n{}^{0,0,0} = \frac{1}{3n+1}\left(\frac{1\cdot 3\cdot\ldots\cdot(n-1)}{2\cdot 4\cdot\ldots\cdot n}\right)^3 \frac{2\cdot 4\cdot\ldots\cdot(3n)}{1\cdot 3\cdot\ldots\cdot(3n-1)}$$

$$= \frac{1}{3n+1}\left(\frac{n!}{(n/2)!}\right)^3 \frac{(3n/2)!}{(3n)!}$$

(see HOBSON (1931) p. 86).

For $n=4$ there are nine equations, and for $n=6$, 17. It is possible to check the results by evaluating $E_n{}^{n,n,0}$ using the polynomial and the Beta function, and also by means of the relation

$$\sum_{-n}^{n} E_n{}^{i,i,0} = 0.$$

OUTLINE OF DEVELOPMENT OF THE DAISY

The theory developed in this paper is limited by a number of assumptions which are by no means always satisfied. Two are of special importance:

(i) That the pattern passes through a long developmental period without forming any visible structures, and indeed without the chemical patterns modifying in any way the geometry of the system. When the visible structures are finally formed, this is done without essential alteration of the chemical pattern.

(ii) That the pattern is always developed within a ring so narrow that it may reasonably be treated as a portion of a cylinder.

The first of these assumptions is one which it would be very difficult to avoid. It would be exceedingly difficult to know what to assume about the anatomical changes. For the majority of plants this assumption is probably false. In the development of the capitulum of a daisy it seems to be more or less correct, however. The capitulum is appreciably separated from the rest of the plant by a length of petiole before the development of the capitulum starts. Thus a new start is made in the development of the capitulum. It is not appreciably influenced by the proximal structures. That this is the case is confirmed by the following facts:

(a) The direction of the generating spirals of the rosette and of the capitulum are statistically independent. Thus of fifteen capitula and corresponding rosettes examined by the author, four cases had both rosette and capitulum left handed. In five cases the rosette was left handed but the capitulum right handed, and in four the rosette right handed and the capitulum left handed; in one case both were right handed. Thus in nine out of the fifteen cases the rosette and the capitulum were in different directions.

(b) Beneath the thirteen bracts enclosing the capitulum there are no other distinguishable structures.

It is suggested that the development of the daisy proceeds essentially as follows. First, the petiole grows up from the rosette without any differentiation either of a visible anatomical form or of an invisible chemical form. Subsequently the distal end of the petiole undergoes two kinds of change. Its diameter increases and at the same time a chemical pattern is determined by purely chemical considerations, and there is therefore little reason to expect the wavelength to change much. As the diameter increases further, therefore, the pattern will have to change in order that it may continue to fit on the petiole with its new diameter. A very rough description of the concentration patterns during this process may be described as follows:

The concentration U of one of the morphogen concentrations $x = (\varrho, \theta, z)$ is to be given by the formula

$$U = \sum_{\eta} e^{i(\eta, x)} G(\eta^2) W(x) \tag{I}$$

where the summation is to be over the lattice

$$A = \begin{pmatrix} A & B \\ C & D \end{pmatrix}$$

reciprocal to

$$a = \begin{pmatrix} a & b \\ c & d \end{pmatrix}.$$

The precise form of the function $G(\eta^2)$ is not known, but it would be suitable to construct it to fall from an initially high value for $\eta^2 = 0$, rise to a maximum near the square of the shortest vector of the lattice A, and then tend to zero for large η^2. The function $W(x)$ should depend only on z and typically may be of the form $\exp(-z^2/2\sigma^2)$. The ratio of the standard deviation σ to the shortest vectors of the lattice a is probably between two and five. The inclusion of this factor $W(x)$ of course results in the pattern not having the symmetry of the lattice a, or of any other lattice. But it is nevertheless possible to use the lattice a applying to the formula (I) to describe the pattern instead of the symmetry lattice. It remains only then to describe what in the lattice is to be used for each value of the diameter of the petiole. A suitable form for the lattice is the limiting divergence angle lattice described in Part I. Clearly this description cannot hold at all times. It breaks down for the period during which the pattern is beginning to form. There may also be a period during which there is a pattern with reflexion symmetry (e.g. a decussate pattern), and the formula above will be invalid for this period also. The sections which follow are concerned with considering the chemical conditions under which this sort of description of the pattern very broadly holds.

At a certain point in the development of the daisy the anatomical changes begin. From this point, as has been mentioned, it becomes hopelessly impracticable to follow the process mathematically; nevertheless it will be as well to describe how the process does proceed (at least in the author's opinion). In the regions of high concentration of one of the morphogens, growth is accelerated, and subsequently florets appear. Also the chemical pattern begins to spread inwards towards the apex, and the florets follow it. The wave length of course remains essentially unaltered during this inward movement and therefore as the apex is approached the parastichy numbers fall, producing the usual disc pattern, possibly with some slight irregularity at the very centre. There may still be some growth of the capitulum itself, but the pattern can no longer adjust itself to keep the wavelength constant. Either the chemical pattern has lost all its importance and gives way to the relatively unchangeable anatomical pattern, or else secretions from the new structures ensure that the wavelength of the chemical pattern increases with that of the anatomical pattern.

A special point arises in connection with the daisy, the formation of the ring of thirteen bracts. This number is very constant. The author does not recall finding any specimen with a different number of bracts, excepting a very few deformed or damaged specimens. It is suggested that this ring of bracts is formed as follows. Within the band of lattice pattern there appears at some stage a ring of reduced activity, so that the band becomes divided into two separate bands. The more distal of these bands continues its development and eventually forms the floret pattern. The proximal band, however, is rather narrow and weak (it is pointless to enquire why). This process is described in .... The number of maxima in the ring under these circumstances will be one of the three principal parastichy numbers, usually the largest of the three. In view of the fact that the daisy develops according to the normal Fibonacci pattern, this number must be expected to be a Fibonacci number, as it is.

Considerations governing the choice of parameters

The assumptions to be made concerning the development of the pattern are

(i) That the pattern is described by functions U, V of position on the cylinder and of time, satisfying the partial differential equations

$$\frac{\partial U}{\partial t} = \phi(\nabla^2)U + I(x,t)U + GU^2 - 1 + UV,$$

$$V = \psi(\nabla^2)U^2.$$

(ii) The operator $\phi(\nabla^2)$ is supposed to take the form

$$\phi(\nabla^2) = I_2\left(1 + \frac{\nabla^2}{k_0^2}\right)^2.$$

(iii) The operator $\psi(\nabla^2)$ is supposed to take the form

$$\psi(\nabla^2) = \frac{1}{1 - \nabla^2/k_0^2}$$

though in the computations other forms may be used, taking the value zero outside a finite region.

(iv) A quasi-steady state is assumed to hold, i.e. the time derivative $\partial U/\partial t$ is supposed to be zero, or so near zero as is consistent with slow changes in the radius of the cylinder. This assumption of course implies that certain details as to the effect of the growth on the equation need not be considered.

(v) The function $I(x, t)$ is supposed given in advance. At each time it may be supposed to take the form $I_0 - I_2 z^2/l^2$. The quantity I_0 is initially supposed to be negative and to increase to an asymptotic value, reaching very near to it when the optimum wavelength is about one third of a circumference. The quantity l can remain very nearly constant or increase slightly with increasing radius. Clearly in view of (iv) it is only the variation of I_0 and l with radius which is significant, not the variation with time.

If we concentrate our attention on the period of time in which the optimum wavelength is less than a third of a circumference, I_0 and l may be taken as constants, i.e. on a par with G, H, I_2, k_0, R. We have to consider what are appropriate values for these seven quantities. Of the seven quantities there are really only four that are dimensionless. In other words, if we are quite uninterested in the units of time, length and concentration, new units may be introduced which will result in three of these parameters taking the value unity. Actually it is not advisable to do this reduction in every context. A certain amount of interest attaches to the relation of the time and space scales of the phenomena and the diffusion constants for the morphogens in the tissue. The enormous variety of possible reaction constants, and the fact that exceedingly weak concentrations of morphogens could be effective to influence growth, mean that our ignorance of the other two dimensionful quantities is too great for there to be any value in considering them in detail.

If three of the parameters are to be taken as unity, appropriate ones seem to be k_0, fixing the unit of length as the optimum radian wavelength, I_2, fixing the unit of time, and G, fixing the unit of concentration.

The choice of the parameters H, I_0, l is assisted by obtaining an approximate form of solution valid for patterns covering a large area, i.e. in effect with l very large. One may then, as a very crude approximation, suppose that when $I(x, t)$ varies from place to place one may find near each point more or less the solution which would apply over the whole plane if the value of I appropriate for that point were applied to the whole plane. A nomogram for this purpose is given elsewhere. [This appears to have been lost.] Another approach to the problem is provided by considering the effect of the terms $\phi(\nabla^2)U$ and $I(x, t)U$ taken in conjunction in the absence of the terms $GU^2 - HUV$. The terms $I(x,t)U$ may then be regarded as modifying the effect of the $\phi(\nabla^2)U$ term, so that $\phi(\nabla^2)$ has to be replaced by another function of the wave vector, no longer dependent on the length alone. Having expressed the effect of the $I(x,t)$ term in this way, it may be assumed, as another (alternative) crude approximation, that the effect of this term is the same even in the presence of the terms $GU^2 - HUV$. Clearly this approximation will not be too unreasonable if the really important term is $\phi(\nabla^2)U$.

Another way of expressing the effect is that the poison, acting through the HUV term, prevents the growth of waves whose wave vectors are near to that of a strong wave train. The quantity R expresses essentially the range of action in the wave-vector space. If it is too small, there will be liberty for "side bands" to develop round the strong components. These side bands will represent the modulation of the patchiness. If R is allowed to become too large, it can happen that this "side band surpression" effect even prevents the formation of a hexagonal lattice; neighbouring points around the hexagon of wave-vectors surpress one another. This however happens only with certain values of the other parameters.

In the actual calculations (initially, at any rate) the function chosen for $\psi(\nabla^2)$ was

$$\psi(r^2) = \begin{cases} \left(1 - \left(\dfrac{r}{r_{max}}\right)^2\right)^2, & r \leqslant r_{max}, \\ 0, & r \geqslant r_{max} \end{cases}$$

with r_{max}/k_0 usually about $1/\sqrt{2}$.

BIBLIOGRAPHY

ADLER, I.
1974   A model of contact pressure in phyllotaxis
        J. Theor. Biol. **45**, 1–79.

ARBER, A.
1950   *The Natural Philosophy of Plant Form*
        (Cambridge University Press, Cambridge).

BRAUN, A.
1835   *Flora*, 145.

BRAVAIS, L. and A. BRAVAIS
1837   *Ann. Sci. Nat. Botanique* (2) **7**, 42 and **8**, 193.

BURTON, D.M.
1976   *Elementary Number Theory*
        (Allyn and Bacon, London).

CASTETS, V., E. DULOS, J. BOISSONADE and P. KEPPER
1990   Experimental evidence of a sustained standing Turing type non-
        equilibrium chemical pattern
        Phys. Rev. Lett. **64**, 2953–2956.

CHURCH, A.H.
1904   *On the Relation of Phyllotaxis to Mechanical Laws*
        (Williams & Norgate, London).

CHURCH, A.H.
1920   *Oxf. Bot. Mem.*, No. 6.

DAWKINS, R.
1986   *The Blind Watchmaker*
        (Longmans, London).

DE BEER, G.
1951   *Embryos and Ancestors*
        (Oxford University Press, Oxford).

GOEBEL, K.
1922   Gesetzmässigkeiten in Blattaufbau
        Bot. Abh. (Jena) **1**.

HARLAND, S.C.
1936   The genetical conception of the species
        Biol. Rev. **11**, 83.

HOBSON, E.W.
1931   *The Theory of Spherical and Ellipsoidal Harmonics*
        (Cambridge University Press, Cambridge).

HODGES, A.
1983   *Alan Turing: The Enigma of Intelligence*
        (Hutchinson, London).

MURRAY, J.D.
1981   A pre-pattern mechanism for animal coat markings
        J. Theor. Biol. **88**, 161–199.

PALEY, W.
1802   *Natural Theology*
        (F. & J. Rivington, London).

RICHARDS, F.J.
1948   The geometry of phyllotaxis and its origin
        Symp. Soc. Exp. Biol. **2**, 217–245.

ROSS-CRAIG, S.
1951   *Drawings of British Plants, Part V*
        (Bell and Sons, London).

THODAY, D.
1939   The interpretation of plant structure
        Nature **144**, 571.

THOM, R.
1972   *Stabilité Structurelle et Morphogénèse*
        (Benjamin, Reading).

THOMPSON, D' A.W.
1917   *On Growth and Form*
(Cambridge University Press, Cambridge).

VAN ITERSON, G.
1907   *Mathematische und Mikroskopische-Anatomische Studien der Blattstellungen*
(Fischer, Jena).

WARDLAW, C.W.
1953   A commentary on Turing's diffusion-reaction theory of morphogenisis
New Phytol. **52**, 40–47. Republished in *Essays on Form in Plants*
(Manchester University Press, Manchester, 1968).

INDEX

[130]

Printed and bound by CPI Group (UK) Ltd, Croydon, CR0 4YY

03/10/2024

01040329-0014